Violent Phenomena in the Universe

Jayant V. Narlikar

Dover Publications, Inc.
Mineola, New York

Bibliographical Note

This Dover edition, first published in 2007, is an unabridged republication
of the work first published by Oxford University Press, Oxford, 1982.

Library of Congress Cataloging-in-Publication Data

Narlikar, Jayant Vishnu, 1938–
    Violent phenomena in the universe / Jayant V. Narlikar. — Dover ed.
      p. cm.
    Originally published: Oxford ; New York : Oxford University Press,
1982.
    Includes bibliographical references and index.
    ISBN 0-486-45797-4
    1. Astrophysics. 2. Cosmology. I. Title.

QB461.N34 2007
523.01—dc22

2006102949

Manufactured in the United States of America
Dover Publications, Inc., 31 East 2nd Street, Mineola, N.Y. 11501

Preface

During the 1960s and the 1970s the subject of astronomy was considerably enlivened by inputs from observations made from above the Earth's atmosphere. Balloons, rockets, and satellites have enabled man to probe the distant parts of the Universe with those wavelengths of the electromagnetic radiation which were hitherto not available to him on the surface of the Earth because of their absorption by the Earth's atmosphere. Of particular significance in all these developments has been the spectacular rise of X-ray astronomy, culminating in the launching of the Einstein Observatory. The wealth of data from this X-ray satellite has, in a short time, given considerable food for thought to the theoreticians in their attempts to understand the nature of the Universe.

These remarks, of course, do not take away any credit from the well established branches of ground-based astronomy – from the age-old optical astronomy and the post World War II radio astronomy. Indeed the two most remarkable discoveries, of quasars in 1963 and of pulsars in 1968, belong to these two disciplines. The sophistications of modern electronics have found their way into the various devices that make the modern optical and radio telescopes so much superior to their counterparts a generation ago.

What is the picture of the Universe that is emerging out of these and other branches of astronomy? The one aspect of our Universe that strikes the astronomer most is that the cosmos is not altogether as quiet and peaceful as the clear night sky would make us believe. There are many types of astronomical objects which are now being seen as seats of violent activity – objects where large quantities of energy are being poured out, at rates far beyond any that could be imagined from terrestrial catastrophes. There are objects which have been emitting radiation over millions of years and there are objects which are active over a matter of a few seconds. The radiation is in different forms of electromagnetic waves, ranging from radio waves to gamma rays.

Naturally theoreticians have been kept on their toes by these developments; and they have responded to the challenge of explaining these weird phenomena by constructing ingenious theories. The purpose of this book is to present to the reader an account of some of the highlights of this violent Universe and of the attempts by the astrophysicists to make sense of these observations.

On the theoretical side, the decade of the seventies may rightly be called the 'decade of the black hole'. During this period the general relativists studied and elucidated the mathematical structure of these esoteric objects. During this period the astrophysicist came to look upon the black hole as the cure to all his headaches in trying to resolve the mysteries of the violent Universe. During this period the astronomer came close to making the claim that black holes exist in X-ray binaries, and may be in globular clusters and in the nuclei of galaxies.

To do justice to the subject I have discussed the properties of gravity in the early chapters: for, of all the known interactions of basic physics, gravity must hold the key to the mystery of these cosmic energy machines. Black holes are introduced only after the more familiar aspects of gravity have been presented. In the later chapters the violent phenomena are discussed in their various manifestations, as supernovae, as pulsars, as X-ray sources, as active galaxies, as radio sources, and as quasars. The list is not exhaustive, but is representative enough to indicate to the reader what the modern subject of high energy astrophysics is all about.

The final two chapters deal with the big bang, the violent event through which the Universe was born. The first of these chapters is along conventional lines describing how the physicists have been pushing their theoretical discussions closer and closer to the epoch of the big bang. Some of these attempts are exciting if only because they bring cosmology and particle physics close to each other.

I would have ended the book at Chapter 8 had I shared the optimism of many of my colleagues: an optimism that the big bang picture based on Einstein's general relativity is an essentially correct one. The recent observations of the discrepant red-shifts of quasars may well force us to do some radical rethinking about the nature of gravity. I have deliberately chosen to end the book along unconventional lines to emphasize my personal reservations about whether the problem of the origin of the Universe – the fundamental problem of cosmology – has been solved.

As in my earlier OPUS book *The Structure of the Universe* I have used a few boxes to elucidate further some of the technical aspects of the subject being discussed in the main text. I have also used appendices at the end to discuss a few topics referred to frequently in the main text. A glossary and a few tables of reference matter included at the end of the book may also prove useful to the reader.

I gratefully acknowledge the help of my colleagues Kumar Chitre, Krishna Apparao, and S. Ramadurai and my father Professor V. V.

Narlikar who looked through parts of the manuscript and suggested important changes which led to its improvement. The staff from the photography section and the drawing office of T.I.F.R. were of great help in the preparation of the figures. I also thank Mr. D. B. Sawant for typing the manuscript.

JAYANT NARLIKAR

Tata Institute of Fundamental Research
Bombay, India

Contents

Note on Illustrations   x

1 The Violent Universe   1

2 Gravity According to Newton   14

3 Gravity According to Einstein   37

4 The Physics of Black Holes   52

5 Star Explosions and their Aftermaths   76

6 Powerful X-ray Sources in our Galaxy   97

7 Active Galaxies and Quasars   119

8 The Big Bang   139

9 The Big Bang Revisited   163

Appendix A The Electromagnetic Wave and the Photon   185

Appendix B The Special Theory of Relativity   188

Appendix C The Doppler Effect   193

Appendix D Sub-atomic Particles   195

Appendix E Types of Electromagnetic Radiation   198

Table I Catalogues of Astronomical Objects   202

Table II Fundamental Constants of Physics and Astrophysics   204

Glossary   205

Further Reading   212

Index   213

Note on Illustrations

Most Figures appear in the text, near where they are referred to. But the following Figures, which are half-tones, are gathered together into two groups of plates near the middle of the book:

1.1	1.9
1.2	5.1
1.5	7.1
1.7	7.2
1.8	

1 The Violent Universe

GURU: Today I will discourse upon the violence in astronomy.

DISCIPLE: Revered Sir! Will you be describing the violent phenomena in the Universe?

GURU: Yes, and I will also dwell upon the controversies amongst the astronomers about what these events imply – controversies which are no less violent than the phenomena themselves.

The one aspect of the star-studded night sky which impresses the casual observer most is its tranquillity. The peace and quiet of the heavens with their marked contrast to the hurly burly of life on the Earth have inspired poets, philosophers, and religious thinkers from time immemorial. Even the amateur astronomer viewing the night sky from his proverbial roof top telescope sees a picture which changes very slowly from night to night. The occasional visit of a comet, the fall of a meteorite, or in these modern times the passing of a man-made satellite are examples of events which introduce transitory variations on an apparently steady cosmic theme.

But appearances can be deceptive! What may appear peaceful and steady to the casual observer in fact hides many turbulent phenomena. The observational and theoretical advances in astronomy since World War II have revealed the existence of many types of violent events in the cosmos, events which by their sheer magnificence far surpass any spectacular happening on the Earth. This book is concerned with a description of such violent events in astronomy.

How is this violence in astronomy revealed to the astronomer? How can he estimate the physical power behind any particular violent event? What explanations can he offer for the *cause* of the event? Before we concern ourselves with such questions let us take a look first at a sample of the events themselves; events which point to some violent activity in astronomical objects.

The Crab Nebula

Figure 1.1* shows a photograph of an astronomical object in the constellation of Taurus. This object is not visible to the naked eye or even

* See note opposite for position of Figures.

through an amateur's telescope. Its picture has been obtained by exposing a photographic plate for an extended period at one of the leading astronomical observatories.

The filamentary structure of the bright object gives it a crab-like appearance. To the astronomer, it also indicates some violent activity at the centre. This indication is borne out by a detailed examination of the nebula. For example, velocity measurements indicate that the bright matter in the object is moving away from the centre with speeds around 1000 km per second or more. The more recent investigations by radio and X-ray techniques show that the Crab Nebula is also a source of radio waves, X-rays, and gamma rays.*

This remarkable object is believed to be the remnant left when a star exploded many centuries ago. To be precise, the explosion of the star was first seen on the Earth on 4 July 1054. How are we able to pinpoint the date so accurately? The reason for fixing on this date lies in the fortunate circumstance that the Chinese and the Japanese astronomers of the 11th century recorded this event in their chronicles, most probably for astrological purposes! According to these records the star was so bright in the initial stages that it could be seen in the day time. Even by 27 July 1054 the star was as bright as Venus. Subsequently it faded and by 17 April 1056 it was no longer visible to the naked eye.

It would be interesting to speculate why this object was not seen (or recorded) by observers in Europe and India, where it *should* have been visible. Did the intellectuals of medieval Europe, so dominated by the Christian dogma that God created the Universe in its entire perfection, fail to find a proper place in the scheme of things for the sudden appearance of a strange star? Could the astrologers of India have missed seeing the object, as July is a month of Monsoon rains?

Speculations apart, there are less direct indications that in two other parts of the world this remarkable object was seen. One indication is on the stone pictographs found in the caves of the Pueblo Indians of North America, showing a crescent Moon and a nearby bright object. The position of the object in relation to the Moon seems to agree with that of the exploding star in the Crab Nebula. These pictographs suggest that the Red Indian tribe in this area was sufficiently impressed by the event to have felt the need to record it on rocks. More recently Kenneth Brecher, Elinor Lieber, and Alfred Lieber have pointed out that the Arabs also noticed and recorded the event. A report by Ibn Buttan, a Christian physician of Baghdad who lived in Cairo until late 1052 or early 1053 and later spent a year in Constantinople, states the appearance of a spectacular star in

---

* For a brief description of these different aspects of the electromagnetic radiation see Appendix A.

Gemini some time between 12 April 1054 and 1 April 1055. (Because of the precession of equinoxes this position in Gemini corresponds to the present p asition of Crab in Taurus.)

An exploding star of the type which left the remnant in the form of the Crab Nebula is known as a *supernova*. Since 1054 two more supernovae have been seen to explode in our Galaxy; one by Tycho Brahe in 1572 and the other by Johannes Kepler in 1604. More supernova explosions have been observed in other galaxies. What is the cause of these explosions?

Pulsars

Does a supernova explosion blow apart the entire star? As we shall see later, there are reasons to believe that the central core of the star may survive the explosion. If so, in what way can we expect to see it? While theoreticians speculated about this question, an unexpected observational discovery in 1968 provided the most plausible answer.

Jocelyn Bell, a graduate student in the Cavendish Laboratory at Cambridge University, was making measurements of interplanetary scintillations, with the help of a large radio telescope. Apart from the expected pattern of radiation, she also detected another, rather unusual pattern. This pattern was remarkable for two reasons: it was highly regular and it was of the very short period (in seconds) of 1.337 279 5 ± .000 002 0. The fact that the period can be quoted to seven places of decimal indicates how regular the pattern of pulses was. The small time scale was remarkable since no astronomical objects were then known to show a pattern of radiation with such a rapid variation (see Fig. 1.2).

Jocelyn Bell, her supervisor Antony Hewish, and some of their colleagues at the Mullard Radioastronomy Observatory at the Cavendish Laboratory investigated this unusual pulse pattern. They ruled out the possibility that the pulses could have originated in a planet going round a star. And with this conclusion went away the exciting possibility that these signals were being sent by an advanced civilization of 'little green men'! Instead, these radioastronomers came to the conclusion that the pulses originated in a compact astronomical source which was named a *pulsar*.

Although the first pulsar, now known as CP-1919 (CP stands for the Cambridge Catalogue of Pulsars), was detected by accident, its radiation characteristics were unusual enough to inspire radio astronomers all over the world to search for and find other pulsars. There was considerable excitement when a pulsar was detected in the Crab Nebula: for this discovery seemed to provide an answer to the question, raised earlier, about the remnant of a supernova explosion. To date, the number of pulsars exceeds 300 and the astronomers have succeeded in resolving the mystery of the regular short period pulses from these strange objects.

Cygnus X-1

Let us now turn to a young branch of astronomy, the branch which uses X-rays to detect cosmic sources of radiation. The direct observation of X-rays from outer space became feasible only after the dawn of the space age. For the Earth's atmosphere absorbs X-rays from outer space and thereby prevents their detection by ground-based instruments. When it became possible to launch artificial satellites well above the X-ray absorbing layers of the atmosphere, X-ray astronomy came into its own.

A major advance in X-ray astronomy was the launching of the satellite UHURU. This satellite was launched from the east coast of Kenya on 12 December 1970, on the seventh anniversary of its independence. Appropriately, the name UHURU of the satellite means 'freedom' in Swahili, the national language of Kenya.

The UHURU satellite carried an X-ray detector and, although earlier observations had revealed the existence of a few isolated X-ray sources, it was the first time that astronomers were able to get a long list of cosmic X-ray sources. The X-radiation came from many types of sources. Of these we will briefly look at one source which has generated considerable excitement.

Cygnus X-1 is a source in the constellation of Cygnus. The X-radiation from this source showed a periodicity of 5.6 days. That is, the radiation went through one cycle of maximum and minimum in this period. When astronomers examined their photographic plates for the region where Cygnus X-1 was detected they found, very close to its location, a large star of the type known as *supergiant*. From observations of this star using the visible spectrum it became clear that the star is not isolated but that it is a part of a *binary* system. The scenario is illustrated in Fig. 1.3.

Here we see *two* stars going round each other. Of these the star A is the supergiant star mentioned above. What is star B? The star B is not visible, but its presence can be inferred from the gravitational pull it exerts on its companion A. The X-ray emission of Cygnus X-1 seems to be coming from the vicinity of where star B should be located.

Cygnus X-1 is the most dramatic example so far of X-ray sources associated with binary stars. Why it is so dramatic we will see later.

Gamma Ray Bursts

An even younger branch of astronomy, younger than the X-ray astronomy described above, is the astronomy of gamma rays. Like X-rays, gamma rays also get absorbed by the Earth's atmosphere and hence have to be

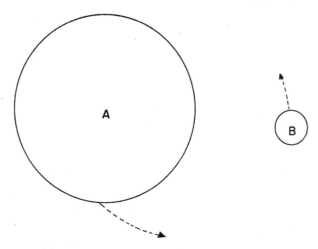

Fig. 1.3. A typical binary system is shown here. The stars A and B go round each other. In the case of Cygnus X-1, the star A is seen to be a big (supergiant) star, but the star B is invisible.

detected with the help of satellite-based detectors. The gamma ray detectors have not yet reached the same level of sensitivity, control, and adaptability as the X-ray detectors; and so gamma ray astronomy is still in its infancy.

Nevertheless the early observations of gamma ray astronomy hold out rich promise of things to come as the detector technology improves. Already astronomers have begun to detect what are known as *gamma ray bursts*.

In a typical burst shown in Fig. 1.4, a large quantity of gamma rays is released in a short time of the order of a few seconds. The burst seems to be a once-and-for-all event; it is not repeated. What type of violent activity may be responsible for a gamma ray burst?

Globular Clusters

In Fig. 1.5 we see a somewhat larger system than just a star or two which we have so far considered. This is a globular cluster, a cluster of stars held together by their gravitational attraction. A cluster like the one shown here may contain as many as a hundred thousand stars.

Notice how the density of stars builds up towards the centre. This is characteristic of a gravitationally bound system, whether it be a single star or a cloud of gas, or a globular cluster. The number of stars increases towards the centre so much that, as seen in Fig. 1.5, it becomes difficult to make out individual star images.

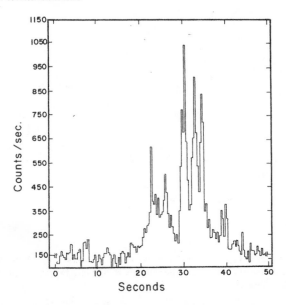

Fig. 1.4. The pattern of radiation from a gamma ray burst is shown here schematically. Notice how rapidly the output of energy rises and falls. This short time scale for which the source is active gives it the characteristics of a burst. This burst was recorded by Apollo 16 in Vela on 27 April 1972. (Based on the work of A. E. Metzget *et al.*, *Astrophysical Journal Letters*, **194**, L 19.)

The discovery of X-ray sources like those detected by the UHURU satellite has added a new dimension to the study of globular clusters, because it is found that some of the X-ray sources are located near the centres of globular clusters. For example, the cluster NGC 6624 (NGC stands for the New General Catalogue) houses the source whose UHURU catalogue number is 3U 1820–30. Is the process of X-ray emission due to the activity going on in the central region of the globular cluster? If so, what is the nature of this activity?

Nuclei of Galaxies

We now consider even bigger systems than globular clusters. Our Galaxy, shown schematically in Fig. 1.6, is a disc shaped object with a central bulge, the entire system containing some hundred thousand million stars. As in the case of globular clusters, the Galaxy also contains an increasingly high density of stars towards its centre. The dynamic activity in the crowded region of the galactic centre gives way to more steady and systematic motions as we move away from it. We are fortunate that our Solar System,

Fig. 1.6. A schematic picture of our Galaxy. The Sun is located about two thirds of the way out in the position indicated by a cross.

the Sun with its planets and satellites, is located two thirds of the way out from the centre (see Fig. 1.6). It is doubtful if the Solar System would have survived intact in the central part of the Galaxy.

But, compared to the nuclear regions of many other galaxies, the centre of our Galaxy is a quiet place! In Figs. 1.7 and 1.8 we see two examples of galaxies whose nuclear regions show violent activity. The galaxy shown in Fig. 1.7, NGC 1068, is a galaxy belonging to a special class selected by C. Seyfert for their bright nuclear regions. Compared to the rest of the galaxy, the nuclear region of a Seyfert galaxy stands out in various ways apart from its brightness. There is explosive activity in the nuclear region as well as emission of X-rays. Are these two phenomena related?

Figure 1.8 shows another extraordinary galaxy, known as M 87 (M stands for the Messier Catalogue). There is a jet-like structure emanating from the centre. Whether it is a continuous jet or a succession of blobs of gas ejected by the centre of the galaxy is not yet known. Recent detailed studies of the nuclear regions of M 87 have led to some bizarre theoretical interpretations which we shall encounter in a later chapter.

Extragalactic Radio Sources

In 1946 J. S. Hey, S. J. Parsons, and J. W. Phillips discovered radio waves coming from the direction of the Cygnus constellation. The techniques of radio measurements in 1946 were not accurate enough to pinpoint the location of the source exactly. In 1951 F. G. Smith at Cambridge was able to achieve sufficient accuracy in locating this source to enable optical astronomers to institute a search for a source of visual light in the same place. The radio source was called Cygnus A.

Walter Baade at the Mt Wilson and Palomar (later renamed the Hale) Observatories did find an interesting object at the location of Cygnus A. Figure 1.9 shows the photograph of this object. This is the photograph of a *radio galaxy*. Baade in fact thought that the photograph shows *two* galaxies in collision.

It is interesting to recall a bet which Baade made with Rudolf Minkowski, another leading astronomer at the Mt Wilson and Palomar Observatories. The bet arose when, at the end of a seminar talk on Cygnus

A, Minkowski made sceptical comments about the collision hypothesis, which had been proposed by Baade and Lyman Spitzer to account for the radio emission from Cygnus A. Baade was, however, confident enough to bet on his theory to the tune of one thousand dollars, but Minkowski talked him down to just a bottle of whisky! It was agreed by both sides that the evidence of emission lines in the spectrum of the gas in the source should be taken as a confirmation of the idea that colliding galaxies were involved. A few months later this evidence was obtained and Minkowski conceded the bet. However, Baade later complained that Minkowski himself finished the whisky that he had given in settlement of the bet!

Later events showed that Minkowski was right in consuming the whisky, for subsequent evidence justified his scepticism of the collision theory. Now it is realized that Cygnus A *does not* owe its radio emission to the collision of two galaxies. What goes on in Cygnus A is in fact characteristic of what goes on in the majority of radio sources located outside our Galaxy which have been discovered since 1951. The detailed evidence available in such cases points not to a collision, but to an explosion in the central region of the radio source; an explosion which throws out electrically charged particles in opposite directions, as shown in Fig. 1.10. These fast particles proceed a certain distance from the source and then radiate in the presence of the magnetic field in the region. Again we encounter evidence of violent activity. What is the process that leads to the emission of fast particles in a radio source? From where does the source derive its tremendous power?

Quasi-stellar Objects

In the early days of radioastronomy it became clear from the example of Cygnus A that considerable progress in the understanding of radio sources

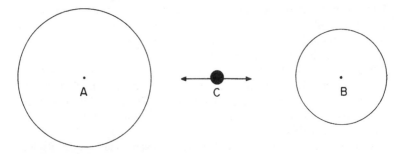

Fig. 1.10. Schematic diagram of a typical extragalactic radio source. The central region C ejects fast particles which radiate radio waves from two blobs A and B located on opposite sides of C.

could be made by their *optical identification*. This process involves locating an object with the help of the optical telescope, in a region close enough to the radio source so that one can argue that the radio object and the optical object relate to the same system. For this process to succeed, the positions of both the objects must be known with good accuracy.

In the early 1960s the occultation of the radio source by the Moon was tried as a means of measuring the position of the source. The Moon's path is known with a great deal of accuracy and the process of occultation, by producing a clearcut drop in the intensity of the source, helps in locating the position of the source behind the Moon. This was the method used in 1962 with the help of the radio telescope at Parkes, Australia by Cyril Hazard. He and his colleagues succeeded in locating precisely the position of the radio source 3C 273 (273rd source in the 3rd Cambridge Catalogue). The optical identification of 3C 273 then became possible and the optical object found in the vicinity of the source had a star-like appearance. In fact, the source was first mistaken for a radio star in our Galaxy. Its extraordinary nature became apparent only when Maarten Schmidt at the Hale Observatories in California examined its spectrum. The spectrum was quite *unlike a normal star* in that it showed a substantial red-shift, and on the basis of his analysis Schmidt concluded that 3C 273 was located way beyond our Galaxy and was at least a *million times* as massive as a typical star such as the Sun. We will return to the property of these red-shifts in Chapters 7 and 9.

This object and another radio source, 3C 48, were the first of the new class of astronomical objects discovered in 1963. Both were star-like in appearance, yet much more massive than stars with spectral peculiarities which placed them very much further away than stars in the Galaxy, and both were emitters of radio waves. These objects were called *quasi-stellar radio sources*, a term subsequently shortened to *quasars*.

Although radio astronomy first led to the discovery of quasars, it soon became clear that not all quasars are radio sources. A number of radio-quiet objects, resembling in other respects the early quasars 3C 273 and 3C 48, were discovered and by now, with more than 1300 quasars known, it is estimated that the property of radio emission may be found in only a few per cent of all quasars.

At the time of writing this book, there is still a controversy about how far away the quasars are located. They may be located very far away, further than the most remote galaxy so far known, or they may be comparatively near by. In either case it is still a mystery as to what power source makes the quasars shine. Whatever this source may be, it is evidently confined in a relatively compact region. Here again, the indications are that some extraordinary explanation is called for to explain the energy production in quasars.

The Big Bang

Perhaps by far the biggest example of violent activity is suggested by the models of the large scale structure of the Universe. The astronomical observations which first suggested these models came in the late 1920s, through the observations of galaxies in the vicinity of our own. Edwin Hubble's analysis of the spectra of these galaxies led him to the remarkable result that most of the galaxies are receding from us with speeds increasing in proportion to their distances. We will examine in Chapter 8 the data which led to this remarkable conclusion.

The mathematical models which are put forward to account for Hubble's observations lead to the concept of the *expanding Universe*. In the expanding universe models, the entire space with the galaxies embedded in it expands so that each galaxy 'sees' the others receding from it (see Fig. 1.11).

If we extrapolate this situation into the past we come to the conclusion that the rate of expansion was *faster* in the past and that the Universe was *denser* in the past than now. How far into the past can we continue our

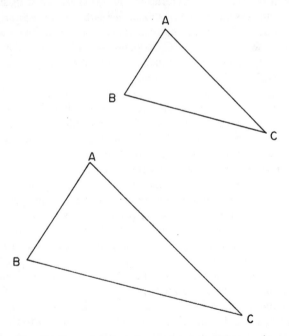

Fig. 1.11. Three galaxies A, B, C are shown at two epochs in an expanding Universe. The triangle formed by A, B, and C has expanded from the earlier to the later epoch. Each galaxy 'sees' the other two as receding.

extrapolation? Conventional wisdom tells us that the extrapolation can be continued until the epoch when the entire Universe was concentrated in a point-like space! This past epoch is estimated to be around ten thousand million years ago. It was at this epoch that the Universe is assumed to have been created and that too in a tremendous explosion. The *big bang*, as the explosion is called, not only marks the beginning of the Universe, it also marks the beginning of all physical concepts including the concepts of space and time. The big bang is, then, the ultimate in all violent phenomena. Whether we like it or not aesthetically, physically, or philosophically, it is to the big bang that we are led if we insist on extrapolating our knowledge of physics to the very remote epochs.

The Role of Gravity in Astronomy and Physics

From the Crab Nebula to the big bang we have now seen various examples of violent activity in the cosmos. The amount of energy involved in such phenomena far exceeds any that can be seen or imagined within the narrow confines of our Earth. For example, the energy stored in a strong radio source like Cygnus A is about ten thousand million million million million million million times the energy released in the explosion of a megaton H-bomb. The natural violent phenomena on the Earth, like earthquakes, volcanoes, hurricanes, and tornadoes, are similarly tiny when compared to the cosmic catastrophes described above.

The detailed mechanism leading to such violent activity is not, of course, the same in all the cases. Yet, the astrophysicist looking for a scientific explanation of the mechanism is led in each case to one basic law of physics which holds the key to the understanding of these violent events: the law of gravity.

To date, the physicist knows four basic interactions which seem to govern the properties of matter. Of these, the *strong* and *weak* interactions operate over very limited ranges and their influence is felt only at the level of particles which form the constituents of atoms: particles like electrons, protons, neutrons, neutrinos, and a large variety of other types whose existence has been detected mainly through machines which accelerate particles to high energies, and through cosmic rays. The *electromagnetic* interaction describes the electrical and magnetic properties of matter and its effect is felt right from the structures of atoms and molecules to the mechanisms which generate radiation in astronomical objects.

The fourth interaction, *gravity*, is very much overshadowed by the above three in determining the behaviour of matter at the microscopic level. The effects of gravity depend on the *masses* of interacting objects: the larger the masses the bigger the effects. At the microscopic level of atoms and molecules these effects are small. The astronomical systems, on the other hand, have huge masses. In our examples above, a star like the Sun has the mass of the order of two thousand million million million million tons. Denoting the mass of the Sun by the symbol $M_\odot$ and using it as a unit, the mass of our Galaxy is around $10^{11}\ M_\odot$ and the mass of visible matter in the Universe lying within the purview of our best telescopes is around $10^{21}\ M_\odot$. It is not surprising, therefore, that gravity plays the key role in determining the equilibrium or otherwise of the astronomical systems.

This is not to deny the relevance of the other three interactions to the astrophysicist! The electromagnetic interaction is of vital importance to the astronomer because it is the electromagnetic wave that brings us information from the heavenly bodies in the first place. In Appendix A we describe the different manifestations of the electromagnetic wave, manifestations ranging from radio waves to gamma rays. Visible light, without which astronomy as a subject would not have started, is also a form of electromagnetic wave. The strong and weak interactions might, at first sight, be dismissed as being irrelevant to the astrophysicist because of their very limited range. Yet, it is now known that these interactions play a crucial role in answering the question 'why do stars shine?' In Appendix D we outline the main features of these basic interactions which are often used by the astronomer. We may have to refer to these features from time to time in the course of this book.

It is gravity, however, which decides which astronomical system is stable, how the constituents of the system should move, and whether the system can lose equilibrium so that it *explodes* or *implodes*. In this book, which is concerned with the violent phenomena in the Universe, our main emphasis will be on gravity.

It is interesting in this connection to review how the role of gravity and the status of astronomy have fluctuated over the years. In the days of Kepler, Galileo, and Newton three to four centuries ago, astronomy occupied pride of place amongst the sciences. It was through astronomical observations that the law of gravitation was proposed by Newton. Gravity was thus the first of the basic interactions to become part of physics. Gradually, however, the importance of gravity dwindled as laboratory physics began to develop. Although astronomical observations kept coming and important discoveries of new planets and satellites were made, the excitement shifted from merely looking at remote objects to studying the properties of matter close at hand. With the interesting and useful developments in electricity and magnetism in the 19th century and with the spectacular progress of atomic and nuclear physics in the early part of the present century, astronomy was relegated to the status of a poor relation. And, in spite of Einstein's creation of a brilliant theory of gravity, the general theory of relativity in 1915, gravity continued to remain neglected and somewhat distant from the other exciting developments in physics.

This downward slide of interest in astronomy and gravity has happily been reversed, thanks to the remarkable observational discoveries of the post World War II era. The realization that the Universe is not just made of steadily evolving stars and the clockwork-like motions of planets and satellites has renewed interest in astronomy as a branch of science and in gravity as a basic interaction of physics. Physicists now appreciate the fact that the newly discovered highly spectacular phenomena in astronomy provide grand testing grounds for the laws of physics. For here these laws can be tested on a scale far beyond the limit ever possible on the Earth. And it is here that the physicist can hope to learn more about gravity, whose mysterious nature has so far defied any attempts at a deeper understanding.

2 Gravity According to Newton

KNELLER: To you the universe is nothing but a clock that an almighty clock maker has wound up and set going for all eternity.

NEWTON: Shall I tell you a secret, Mr. Beautymonger? The clock does not keep time. If it did there would be no further need for the Clock maker . . . Can you, who know everything because you and God are both artists, tell me what is amiss with the perihelion of Mercury?

KNELLER: The what?

NEWTON: The perihelion of Mercury.

KNELLER: I do not know what it is.

NEWTON: I do. But I do not know what is amiss with it. Not until the world finds this out can it do without the Clock maker in the heavens . . .

From: *In Good King Charles's Golden Days* by George Bernard Shaw

In this chapter and the next, we will take a look at some of the basic properties of gravity, properties which are useful to the astrophysicist in his attempts to understand the violent phenomena in the universe. As mentioned in Chapter 1, gravity was the first of the four known basic forces to be discussed within the framework of science. Yet even up to the present day, in spite of the work of geniuses like Newton and Einstein, gravity remains full of mysteries. The physicist today is beginning to get a glimmer of understanding of the scheme which holds together the other three basic forces; but as yet he cannot find a proper place for gravity in such a unification scheme.

In following the properties of gravity we will begin with a historical approach. For in this way we see how the universal nature of the force of gravity was gradually brought home to the astronomer and the physicist. We will keep the mathematical description of the various phenomena to the bare minimum, although a few basic formulae and order of magnitude calculations cannot be avoided.

The Laws of Motion

The first mathematical statement describing the behaviour of gravity as a basic force of physics came from Isaac Newton (1642–1727). In 1687, in his classic book *Philosophiae Naturalis Principia Mathematica* (The Mathematical Principles of Natural Philosophy), Newton succinctly described his ideas on dynamics and gravitation. Before we come to discuss gravitation it is worth while taking a brief diversion to look at Newton's laws of motion, which form the basis of modern dynamics.

The *first* law of motion tells us that a body will continue to move in a straight line with uniform speed unless an external force acts on it. Or in particular, if such a body is initially at rest it will continue to be at rest so long as no force acts on it. This law, known as the *law of inertia*, describes the inherent tendency of the body to maintain the *status quo*. Inertia is the property of the body which characterizes this tendency to resist a change.

In this form the law of inertia was known to Newton's predecessor, Galileo Galilei (1564–1642). In fact it was Galileo who, through a brilliant series of arguments and experiments recorded in his famous book *Dialogues Concerning Two New Sciences*, had demolished the long held dogma which had existed in Europe since the days of Aristotle (384–322 B.C.). This dogma held that force is needed to keep a body moving. Galileo had shown, for example, that if a constant force acts on a body its velocity does not remain constant, as Aristotle's followers believed, but that the application of a constant force causes the velocity of the body to *change* at a constant rate.

It was Newton, however, who gave a quantitative expression to the relationship between the applied force and the change of velocity produced by it. His *second* law of motion is stated in the following form:

force = mass × acceleration.

Here *mass* is a measure of the quantity of matter in the moving body. Newton argued that it is this quantity that provides a quantitative measure of the property of inertia in the body. Acceleration is the rate of change of velocity.

The second law therefore tells us the precise relationship between force, inertia, and acceleration. If the same force acts on two bodies of different masses it will generate a larger acceleration in the body of smaller mass. Conversely, to produce a larger acceleration in the same body we need to apply a larger force.

Let us try to visualize what is meant by acceleration. To do that we first note that the velocity of a body contains two bits of information. It tells us at what speed the body is moving and in what direction it is moving. In Fig. 2.1 we give examples of how the velocity is represented by a diagram.

Here we have the four directions – east, west, north, and south – shown on a horizontal plane. The arrow $\overrightarrow{AB}$ is pointing in the easterly direction, indicating the direction of motion. If we choose a definite scale, the length

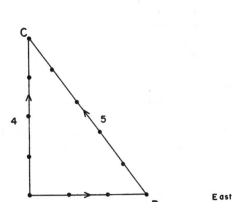

Fig. 2.1. The velocity is a vector which is represented by magnitude and direction. The arrow A⃗B indicates the direction of motion (A to B) and its length (3 units) corresponds to the magnitude of the velocity. The subtraction of velocities is also shown above. The vector B⃗C is the difference between the velocities A⃗C and A⃗B.

of AB can indicate the speed of the body. Suppose AB indicates a speed of 3 metres per second. Then on the same scale the arrow A⃗C in the above diagram shows a body moving with a speed of 4 metres per second in the northerly direction.

Now acceleration measures the rate of change of velocity. The change in velocity may come either from a change in speed, or from a change of direction, or for both these reasons. Suppose in the example given above the velocity of the body initially was indicated by the arrow A⃗B and one second later it was changed to that indicated by A⃗C. What is the acceleration of the body in this case?

To know the answer to this question we must know how to subtract the velocity represented by A⃗B from the velocity represented by A⃗C. Mathematicians call this process the subtraction of *vectors*, vectors being the quantities which have magnitude and direction. Figure 2.1 tells us how to perform the subtraction of the vector A⃗B from the vector A⃗C. The answer is simply the arrow *from* B *to* C. The arrow B⃗C tells us the change in velocity in magnitude and direction. If we measure the length of B⃗C we will find that it corresponds to a speed of 5 metres per second.

Since this change has taken place in unit time, that is in one second, the answer to our question is that the acceleration of the body is 5 metres per second *per second* in the direction from B to C.*

We now come to another example of acceleration which has a bearing on our discussion of gravity. Suppose we have a body moving along a circular curve with uniform speed. Although the speed of the body is the same at all times, it is accelerated because its direction changes continuously. In Box 2.1 we discuss such a motion in more detail. There we

---

Box 2.1 Acceleration in circular motion

In Fig. 2.2 we see two positions of a body moving in a circular track of radius *r*, centred at O. Suppose it is at A at time $t = 0$ and at B at a later time $t = h$, where *h* is small. From A to B, the speed of the body has not changed; it is *v* throughout. However, its direction of motion has changed. The direction of motion at any instant is along the tangent to the circle, as shown by the arrows at A and B.

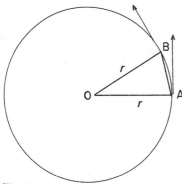

Fig. 2.2

We represent this fact by a velocity diagram where from a point P we draw lines PC and PD parallel to the tangents at A and B respectively. The lengths PC and PD are equal and represent the magnitude of velocity, *v*. Since PC and PD are parallel to the tangents at A and B and these tangents are themselves perpendicular to the radial directions OA and OB, it follows that the angle AOB is equal to angle CPD.

The two triangles AOB and CPD are geometrically similar and hence the ratios of their corresponding sides are equal. Therefore

$$\frac{CD}{AB} = \frac{PC}{OA}.$$

---

* It is customary to indicate the directionality of the segment BC by placing an arrow on top: $\overleftrightarrow{BC}$.

As *h* becomes smaller and smaller, AB very nearly approximates to the distance *h* × *v*. Similarly CD represents the change of velocity, and has a magnitude very nearly equal to *h* × *f* where *f* is the magnitude of acceleration. The above relation therefore gives

$$\frac{h \times f}{h \times v} = \frac{v}{r},$$

that is, $f = v^2/r$. Notice that in the limit as $h \to 0$ our result becomes exact. Also the direction from C to D becomes parallel to the radial direction from A to O. Hence the acceleration of the body is v^2/r, directed towards the centre.

show that if the speed of the body is denoted by the symbol *v* and if the radius of its circular track is *r*, then its *acceleration* is directed towards the centre of the track and has a magnitude given by

$$f = \frac{v^2}{r}.$$

For example, an athlete running at a speed of 6 metres per second along a race track of radius 50 metres has an acceleration given by

$$f = \frac{6 \times 6}{50} = 0.72$$

in units of metres per second per second. The acceleration is directed towards the centre of the track.

The Inverse Square Law

We now leave our discussion of Newton's laws of motion and transfer our attention to Newton's law of gravitation. As stated in Chapter 1, the data which gave shape to this law came from astronomy, from the laws of planetary motion.

After several years of painstaking analysis of the data on planetary positions Johannes Kepler (1571–1630) arrived at the following three laws which seem to govern the motion of planets round the Sun.

The first law tells us that the planets move in elliptical orbits with the Sun as a focus. The second law tells us that a planet moves along its elliptical orbit at such a speed that the line joining the Sun to the planet sweeps out equal areas in equal intervals of time. The third law relates the time *T* taken by the planet to complete one orbit to the *semi-major axis of the ellipse a*, by the following rule: for all planets of the Solar System the ratio $a^3 : T^2$ is the same. In Fig. 2.3 the three laws are illustrated, and the technical terms defined.

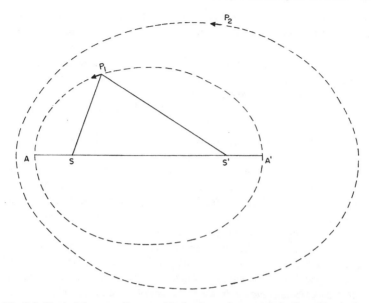

Fig. 2.3. Kepler's laws are illustrated in the above diagram. As indicated by the first law, two typical planets P_1 and P_2 are shown to move on separate ellipses with the Sun S as a common focus. For planet P_1, for example, the second focus is at S'. The line joining S to S' and extended in both directions meets the orbit of P_1 at A A'. The point A is the closest point to the Sun on the orbit while A' is the farthest point. A is called the *perihelion* and A' the aphelion. The length AA' is called the major axis of the ellipse and is denoted by $2a$. The ratio of lengths SS' to AA' is called the eccentricity of the ellipse. Kepler's second law tells us that the line SP_1 sweeps out equal areas in equal intervals of time. The third law tells us that the squares of the times taken by P_1 and P_2 to complete their respective orbits are in proportion to the cubes of their respective major axes. Thus P_2 takes a longer time to complete its orbit than P_1.

Notice how the three laws progressively provide more details of the planetary motion. The first law gives the shape of the orbit, the second law tells us how a planet moves along the orbit, while the third law compares the orbital motions of different planets.

But why do the planets move in elliptical orbits? The fact that a planet does not move in a straight line with uniform speed indicates that there is a force acting on it. This much follows from the first law of motion. What is the magnitude and direction of this force? The answer to this question can be given with the help of Newton's second law of motion. For this purpose we first need to know the magnitude and direction of the *acceleration* of the planet. Once we know the acceleration, the second law tells us that the force on the planet is given by the product of the mass of the planet and its acceleration. Therefore, if we know the acceleration of a planet moving according to Kepler's laws, we can determine the force acting on the planet.

Newton solved this problem in two ways. The longer and the more conventional method that he used was the method of geometry. An example of the geometrical method is seen in Box 2.1 where it is shown that the acceleration of a body moving in a circle with uniform speed is directed towards the centre of the circle. The circle is geometrically a much simpler curve than the ellipse. A geometrical method along the lines of Box 2.1 becomes considerably more elaborate in the case of a planet moving in an elliptical orbit as required by Kepler's laws. Nevertheless Newton could solve this problem and arrived at the result that the acceleration of the planet is directed towards the *focus* of the ellipse where the Sun is located and that the magnitude of the acceleration varies according to the *inverse square* of the planet's distance from the Sun.

Newton's second method was considerably more elegant and direct, but it made use of a new technique of mathematics which Newton himself had invented. He gave it the name of 'fluxions' although it is now commonly known as 'calculus'. How a modern student of dynamics would solve this problem with the help of calculus is illustrated in Box 2.2.

Box 2.2 The inverse square law deduced from Kepler's three laws

Refer to Fig. 2.3 and denote the distance SP_1 by r and the angle P_1SA' by Θ. Then the first law of Kepler tells us that since P_1 moves along an ellipse with focus at S, its equation is given by

$$\frac{1}{r} = \frac{1}{l}(1 - e\cos\Theta)$$

where l is a constant, known as the *semi-latus rectum* of the ellipse. e is the eccentricity.

The radial and transverse accelerations of P_1 are given respectively by

$$f_r = \ddot{r} - r\dot{\Theta}^2; \quad f_t = \frac{1}{r}\frac{1}{dt}(r^2\dot{\Theta}).$$

(Here the overhead dot denotes rate of change.)

Kepler's second law tells us that the rate at which SP_1 sweeps out area is constant. This rate is simply $\frac{1}{2}r^2\dot{\Theta}$. Hence $f_t = 0$ and the force on P_1 must be radial, i.e. along SP_1. To calculate this force we have to evaluate f_r, using the fact that

$$r^2\dot{\Theta} = \text{constant} = h.$$

Since r is known as a function of Θ, this calculation is simple but tedious. It gives

$$f_{\mathrm{r}} = -\frac{h^2}{lr^2}.$$

Finally, let us use Kepler's third law. The area of the ellipse traced by P_1 is given by

$$A = \frac{\pi l^2}{(1-e^2)^{3/2}}.$$

Hence the time taken to trace this area must be

$$T = \frac{2A}{h},$$

since the rate of sweeping of area is $h/2$. We therefore get

$$T = \frac{2\pi l^2}{h(1-e^2)^{3/2}}.$$

The major axis of the ellipse is

$$2a = \frac{2l}{1-e^2}.$$

Kepler's third law states that T^2/a^3 is constant. Hence

$$\frac{T^2}{a^3} = \frac{4\pi^2 l^4}{h^2(1-e^2)^3} \bigg/ \frac{l^3}{(1-e^2)^3} = \frac{4\pi^2 l}{h^2} = \text{constant}.$$

Denoting this constant by $1/K$, we get

$$f_{\mathrm{r}} = -\frac{4\pi^2 K}{r^2} = -\frac{\lambda}{r^2}, \quad \lambda = \text{constant}.$$

This is the inverse square law! We note that the acceleration of P_1 and hence the force on P_1 varies inversely as the square of the distance from S. The minus sign in f_{r} means that the force is directed *towards* S.

Newton in fact first solved the problems of motion with the techniques of calculus and later gave geometrical proofs for the convenience of those of his colleagues who might be unfamiliar with calculus and hence suspect the validity of the results!

We now apply Newton's second law of motion to the result that a planet of mass m_{p} has an acceleration towards the Sun with a magnitude given by the inverse square law:

$$f = \frac{\lambda}{r^2}.$$

Here λ is a constant and the factor r^2 in the denominator shows that the acceleration falls away as the inverse square of the distance r of the planet from the Sun. The second law of motion then gives the *force* on the planet towards the Sun as

$$F_{\mathrm{p}} = m_{\mathrm{p}} \times f_{\mathrm{p}} = \frac{\lambda m_{\mathrm{p}}}{r^2}.$$

Kepler's third law of motion further tells us that the constant λ is the same for all planets. The result derived in Box 2.2 relates λ to the ratio $a^3 : T^2$ which is known to be the same for all orbits:

$$\lambda = \frac{4\pi^2 a^3}{T^2}.$$

Thus a typical planetary orbit is characterized by the constant a (equal to half the major axis of the ellipse) and the orbital period T. As we move outwards from the Sun, we encounter planetary orbits of increasing a and T. However, for all these orbits the above formula gives the same value of λ.

Can we say something more about λ? We can, provided we take into account Newton's *third* law of motion, which we have so far not discussed. This law states that action and reaction are equal and opposite. In other words, if we have two bodies A and B of which A is known to exert a force on B, then B also exerts a force on A which is equal to and opposite to the force of A on B.

In our example of the Sun and the planet, the Sun appears to exert a force of attraction on the planet equal to F_{P} and directed towards itself. We therefore expect an equal force F_{P} to act on the Sun in the *opposite direction*, that is towards the planet. The mathematical expression for the force contains a factor m_{P} (the mass of the planet) in the numerator. We can make it look more symmetrical between the Sun and the planet by writing

$$\lambda = GM_{\odot}$$

so that we have

$$F_{\mathrm{P}} = \frac{GM_{\odot} m_{\mathrm{P}}}{r^2} \equiv F_{\odot \mathrm{P}}.$$

Here we have replaced the suffix P by the double suffix $\odot$ P to show that the force of attraction is symmetrical between the Sun ($\odot$) and the planet (P). Notice that by writing λ as $GM_{\odot}$ we have not violated our earlier conclusion that λ does not change from planet to planet.

Stated in this form, the law of attraction can be generalized to any two bodies A and B of masses m_{A} and m_{B}, separated by the distance r:

$$F_{\mathrm{AB}} = \frac{Gm_{\mathrm{A}} m_{\mathrm{B}}}{r^2}.$$

This is the complete mathematical statement of Newton's law of gravitation. Stated in words, this law tells us that the force of attraction between two bodies varies as the product of their masses and varies inversely as the square of their distance apart. The constant *G* which appears in this law is known as the *constant of gravitation* or the *gravitational constant*.

It is often stated that Newton arrived at this law of gravitation from his encounter with the falling apple while he was sitting in an orchard in his native village of Woolsthorpe in Lincolnshire in the year 1666, when he left Cambridge to avoid the ravages of the great plague. This legend, which first seems to have appeared in the writings of Voltaire, while highlighting Newton's originality and speculative power, hardly does justice to his mathematical and deductive ability. In any case the first complete description of the law as given by Newton appeared some 21 years later in his *Principia*. Considerable discussion has gone on in literature on the history of science about why Newton waited so long before publishing the law of gravitation and whether the genesis of the law rests solely with him. For instance, there are indications that Robert Hooke (1635–1703) had also arrived at the concept of the inverse square law of attraction from the data on planetary motion. We will not go any further into the controversial questions about the genesis of this law. This much is clear: the law would not have been discovered without the input from astronomy, from the data available on planetary motion.

Celestial Mechanics

Astronomy continued to provide evidence in support of the law of gravitation. Many of the developments came after Newton's lifetime. One of these was the appearance of a comet in 1758 as predicted by Newton's contemporary Edmond Halley in 1682.

Halley had noticed that a particular comet was seen in the vicinity of the Sun in the year 1682. Although comets come and go every year Halley discovered that on previous occasions comets of somewhat striking appearance (similar to the comet of 1682) were seen in the years 1456, 1531, and 1607. The comets come near the Sun, go round it, and then recede from the solar neighbourhood. Could the comets seen in the years 1456, 1531, and 1607 be one and the same comet appearing periodically after every 75–6 years? Halley thought so and in terms of Newton's laws of motion and gravitation he could give a reason for his supposition. The comet was moving in an elliptical orbit around the Sun just as any planet would do. The difference between the elliptical orbits of a comet and a planet lies in the circumstance that the latter is nearly circular while the former is highly elongated. In a highly elongated ellipse one end of the major axis lies very close to the focal point where the Sun is situated, while the

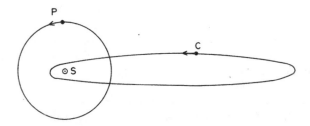

Fig. 2.4. Typical orbits of a comet (C) and a planet (P) are shown here. The cometary orbit is characterized by its large eccentricity (see Fig. 2.3 for definition of eccentricity) while the orbit of the planet is nearly circular. Thus to us on the Earth a comet is seen in the vicinity only when it is near the Sun.

other end of the major axis lies far away from the Sun. The comet is seen only when it is in the part of the orbit close to the Sun and appears to fade away as it recedes towards the far end of the major axis. The situation is illustrated in Fig. 2.4.

Halley's comet takes about 76 years to go round its orbit once. After 1682 its appearance in 1758 therefore caused considerable excitement since the event confirmed Halley's prediction. Like Newton, Halley did not live to see this event. Halley's comet was last seen in 1910 and is next due in the solar neighbourhood in 1986.

It was the French mathematician Pierre Simon de Laplace who really pushed the application of Newtonian theory to the limits then possible. During the years 1799–1825 Laplace solved the complex problem involving the motion of the many planets and their satellites moving in the gravitational influence of one another and of the Sun. A sum total of *eighteen* bodies was involved in this calculation! Laplace's aim was to demonstrate that this complex system moved not at the dictates of a divine agency but according to the laws of dynamics and gravity. Even Newton had felt that the clockwork precision of planetary motion breaks down once in a while and needs to be set right by the Almighty. Laplace's calculations could, however, account for all the observed features of the planetary motion within the framework of the inverse square law.

Laplace's work gave stimulus to the subject now known as *celestial mechanics*. In general this subject deals with the motion of heavenly bodies moving under the law of gravity. Its main applications have been to the Solar System where considerable astronomical data are available to compare with the theoretical predictions. Laplace and his successors evolved elaborate procedures to handle the complicated set of equations whose solutions often involved long series of mathematical terms.

Nowadays these equations are more readily solved on fast electronic computers.

We end this section with a description of two astronomical observations, one a success for Newtonian celestial mechanics, the other a failure.

In 1781 the English astronomer William Herschel discovered the planet Uranus. At that time Uranus was the outermost known planet of our Solar System. However, by 1840 it was becoming noticeable that the orbit of Uranus was not strictly what would be expected on the basis of Newtonian celestial mechanics. The orbit showed departures from the theoretically calculated track.

Whenever there is a disagreement between theory and observation one of two possible deductions can be drawn: (1) the theory is wrong or (2) the observations are incorrect or incomplete. In the case of Uranus two astronomers, J. C. Adams in England and U. J. J. Le Verrier in France, placed their confidence in Newtonian theory and opted for the second deduction. Both independently looked for the cause of the discrepancy in terms of the gravitational pull of a new object in the vicinity of Uranus. Where should this object be located and what should be its estimated mass?

Adams got to the answer first in 1843 and he communicated his conclusions to two leading observers in England, Challis at the observatory in Cambridge and Airy, the Astronomer Royal at Greenwich. Neither Airy nor Challis took much notice of Adams's calculations and did nothing so far as instituting a telescopic search for the object was concerned. Meanwhile, in September 1846 Le Verrier also arrived at similar results which he communicated to J. G. Galle at the Berlin Observatory. Galle promptly looked for the object and found it. It is the planet Neptune, located even further away from the Sun than Uranus.

This episode, apart from illustrating the varying responses of observers to the predictions of theoreticians, is a good example of the second of the two alternatives described above, viz. a discrepancy between theory and observations may imply a new breakthrough on the observational front. We now consider an example of the first alternative. This relates not to a remote planet of the Solar System but to the planet situated closest to the Sun. The discrepancy in this case was not known during Newton's lifetime. Had it been known then it would have caused him a great deal of anxiety!

The planet nearest to the Sun is Mercury, which takes nearly 88 days to make a complete round of the Sun. However, the planet does not seem to be following an ellipse exactly. Its track is illustrated in a somewhat exaggerated fashion in Fig. 2.5. Notice that the orbit is nearly elliptical but at the same time it rotates round the Sun. This is most clearly seen if we look at the point on the orbit situated closest to the Sun. Such a point is called the perihelion. The line pointing from the Sun to the perihelion of Mercury seems to rotate slowly in the sense of the arrow, from one orbit to the next. The actual rate of change is small, about 575″ (arc second) per century. (An arc second is a measure of angles and it corresponds to 3600th part of a degree.) The shift in the perihelion of Mercury is therefore hardly noticeable. Yet it was large enough to worry the theoreticians.

Like the disturbances of the orbit of Uranus, could this progressive shift of Mercury's orbit arise from the gravitational pull of other planets? Certainly the other planets of the Solar System do exert a pull on Mercury significant enough to cause a steady shift of its orbit. Calculations of celestial mechanics which took into account the gravitational influences of all other planets showed that more than 90 per cent of the observed perihelion shift of Mercury could be explained that way.

So out of the original 575″ per century there still remained a small part, 43″ per century, unexplained. This discrepancy was known at the time of Le Verrier who tried to account for it by a similar technique which led to the discovery of Neptune. Le Verrier postulated an innermost planet *Vulcan*, whose gravitational force could have caused the unexplained discrepancy in the motion of Mercury. This planet was looked for but not found. We shall return to this unexplained fact in the following chapter. Small though it is, it seems to demand a radically new approach to gravity, an approach very different from the inverse square law of Newton.

Gravitational Energy

When we lift a heavy suitcase or climb a hill we feel tired because we have to spend energy to do work against the force of gravity. The precise-

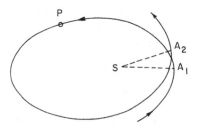

Fig. 2.5. The advance of the perihelion of Mercury shown in an exaggerated fashion. The direction SA, from the Sun to the perihelion, gradually changes from one orbit to the next (shown as going from A_1 to A_2 in the above figure).

relationship between force, work, and energy is not difficult to understand. Suppose a force F is applied in order to displace a body through a distance d along its direction. The force F is assumed to act throughout the process of displacement. Then the work done *by* the force in this process is simply given by

$$W = F \times d.$$

Let us take the example of a body dropped from a height of 10 metres until it hits the ground. The agency which causes the displacement of the body is the force of gravity exerted by the Earth. Newton's inverse square law tells us how to compute the magnitude of this force on the surface of the Earth. For this computation we need the mass of the Earth $M_\oplus$, the mass of the body m, and the distance R between them. Taking the Earth to be perfectly spherical, this distance is simply equal to the radius of the Earth. Let us assume that the mass of the body is $m = 1$ kg, and substitute the values of $M_\oplus = 6 \times 10^{24}$ kg and $R = $ Earth radius $= R_\oplus = 6\,400\,000$ metres. We further need to know the magnitude of G, the gravitational constant. In S.I. units* we have $G = 6.668 \times 10^{-11}$ m^3 s^{-2} kg^{-1}. This means that the force of attraction between two masses of 1 kilogram (kg) each, separated by a distance of 1 metre, is strong enough to generate an acceleration of 6.668×10^{-11} metre per second per second (m s^{-2}) in each mass towards the other. In S.I. units the force which generates an acceleration of 1 metre per second per second in a 1 kg mass is called 1 *Newton*. Therefore the force exerted by the Earth on the 1 kg body is given by

$$F = \frac{Gm\,M_\oplus}{R^2_\oplus} = 9.81 \text{ Newtons.}$$

This force of attraction of the Earth is often called the *weight* of the body. Instead of quoting the above figure for F, we often use the *incorrect* statement that the weight of the body is 1 kg. What we imply is that the weight is equal to the force acting on the body whose *mass* is 1 kg. If the same body were lifted to a height $R_\oplus$ above the surface of the Earth its 'weight' would drop to a quarter of the value on the surface of the Earth, but its mass would not change.

However, the weight of the body would hardly change from its surface value if it were lifted to a height of a few metres. The decrease in the weight would be by about three parts per million at a height of 10 metres. Ignoring such small variations, we may assume that the force of attraction

* These are *Systeme International* units for various physical quantities like length, mass, time, etc. Length is measured in metres, mass in kilograms, and time in seconds.

of the Earth on the body remains more or less the same at the value $F = 9.81$ Newtons computed above throughout its 10 metre fall. The work done by gravity on the body is therefore $F \times d = 98.1$ S.I. units, as it falls through the height of 10 metres. The S.I. unit for measuring energy is *joule* and it is usually denoted by the letter J.

Conversely, this is the work that an external agency must do in order to lift the body from the surface of the Earth to a height of 10 metres. For the process of lifting involves opposing the force of gravity. If the external agency had an energy reservoir which was metered, it would discover that the reservoir is depleted by 98.1 J after the body has been lifted up.

We often hear the phrase 'conservation of energy' which in physical terms implies that energy as a whole cannot be created or destroyed: it can be transformed from one form to another. What has happened to the energy in the above example, to the energy which was spent by the external agency in lifting the body? The law of conservation of energy requires this energy loss to be compensated elsewhere.

The compensation in fact occurs in the form of the gravitational potential energy which the body acquires as a result of its height. The word 'potential' indicates that the source of energy is not explicitly seen but is nevertheless there and can do useful work if required. For example, if the body drops from the height it picks up speed and hence acquires energy of motion, that is, *kinetic energy*. This kinetic energy owes its origin to the gravitational potential energy which the body had by virtue of its height. The best practical example of the use of gravitational energy in this way is that of a hydroelectric power house operated by the gravity dam. In such a dam water falls from a height and its kinetic energy is utilized in driving the dynamos needed to generate electric power. One of the tallest gravity dams in the world is the Bhakra Nangal Dam in the state of Punjab in North India where the water falls through a height of about 200 m. The dam can generate electric power at a rate of 1200 MW (1200 million joules per second).

This is our first example of gravitational energy. Gravity dams tap only a small fraction of the energy available from Earth's gravity. Can we form an estimate of the gravitational energy associated with the Earth as a whole? We can, if we indulge in a thought experiment.

The thought experiment begins with an extrapolation of our earlier calculation of the work done in raising a kilogram of matter through a height of 10 metres. Let us ask: how much work will be needed to raise this bit of matter *through an infinite height*? Our formula for work $W = F \times d$ suggests that this work will turn out to be infinite if we set d equal to infinity. The correct answer, surprisingly, is finite. As shown in Box 2.3,

Box 2.3 How deep is the Earth's gravity well?

Let us calculate, using calculus, the work done in raising unit mass from the surface of the Earth out to *infinite* height.

At a distance r from the centre of the Earth the force of attraction *towards* the Earth on our unit mass is

$$f = \frac{GM_\oplus}{r^2}$$

where $M_\oplus$ is the mass of the Earth. The work done in raising this unit mass through a small height dr is given by

$$f \times dr = \frac{GM_\oplus}{r^2} \cdot dr.$$

To raise the unit mass from the Earth's surface where $r = $ the Earth's radius $= R_\oplus$ (say) to an infinite height, the work required to be done is

$$\int_{R_\oplus}^{\infty} f\, dr = \int_{R_\oplus}^{\infty} \frac{GM}{r^2}\, dr = \frac{GM_\oplus}{R_\oplus}.$$

This quantity gives an estimate of the strength of gravity on the surface of the Earth. In S.I. units this is equal to about 6.3×10^7 joules.

with the help of simple calculus the answer comes out to be

$$W = \frac{GM_\oplus}{R_\oplus} = 6.3 \times 10^7 \text{ J}$$

in our earlier work units. The answer is finite because as we go farther and farther away from the Earth its gravitational pull drops off as required by the inverse square law. Consequently we have to do less and less work to take the body through the same range of height as the body recedes from the Earth. At a height of $R_\oplus$ the work done to raise the body further by 10 metres is a quarter of the work needed to be done to take the body up through 10 metres on the surface of the Earth. At a height of $2R_\oplus$ this fraction falls to 1/9, at $3R_\oplus$ to 1/16 and so on. If we add the progressively decreasing quantities of work all the way to infinity the answer is as given above.

We now ask a further question: 'If the Earth were to be gradually chipped off at the surface and the bits taken to infinite distance, until the entire Earth had vanished, how much work will have to be done?' If we begin chipping off the Earth into 1 kg bits, say, then to start with we would need to do an amount W of work for taking each bit infinitely far away. But in this process both the mass M and the radius R of the Earth begin to decrease. So the final answer will depend on how the ratio $M:R$ changes as the matter from the Earth is removed, that is, on how matter is distributed within the Earth. If the matter were distributed *uniformly* the total amount of work done in stripping the Earth apart turns out to be

$$W_\oplus = \frac{3}{5} \frac{GM_\oplus^2}{R_\oplus}.$$

We do not know the precise density distribution of matter within the Earth and so we cannot determine the correct numerical coefficient in $W_\oplus$. The coefficient 3/5 in $W_\oplus$ above determined on the assumption of a uniform density is not correct but may not be too far wrong. Certainly, so far as an order of magnitude estimate is needed, the above formula for $W_\oplus$ will give us some indication. If we put numbers in the above formula we get

$$W_\oplus = 2.4 \times 10^{39} \text{ ergs}$$
$$= 2.4 \times 10^{32} \text{ J}$$
$$= 6.6 \times 10^{25} \text{ kilowatt hours.}$$

Here 'erg' is a unit of energy: 10^7 ergs make one joule. The kilowatt hour is the basic unit for measuring domestic electricity consumption. This enormous value of $W_\oplus$ in kilowatt hours tells us how tight is the gravitational binding force which holds the Earth together. Only by expending an amount $W_\oplus$ of energy can we tear the Earth apart.

In general, for any spherical astronomical object of mass M and radius R, the quantity

$$W = \frac{3}{5} \frac{GM^2}{R}$$

gives the order of magnitude estimate of its gravitational binding energy. In Table 2.1 below this value is given for a few different astronomical objects. If the object is not spherical a rough estimate of W can be given by substituting for R a characteristic linear size.

Table 2.1 Gravitational Binding Energies

Astronomical object	Binding energy (in joules) estimated by $3GM^2/5R$
Moon	1.2×10^{29}
Earth	2.4×10^{32}
Sun	2.4×10^{41}
Typical white dwarf	2.4×10^{43}
Typical neutron star	10^{46}
Our Galaxy	5×10^{52}

The gravitational binding energy can give us some idea of the energy reservoir in the object. For if there is a significant change of size or shape in the object it may lead to a change of its binding energy and this may release the balance in the form of other types of energy. For example, if the object were to contract to half its original size, the binding energy is doubled. This means that in the process of contraction gravity has done work which results in the gain of energy for the object. This energy may, for example, show up as kinetic energy or in the form of heat.

It was this gravitational reservoir of energy that was first thought to be the likely source of the Sun's brightness. In the last century two physicists, Lord Kelvin in Britain and Hermann Helmholtz in Germany, proposed this idea. As we see from Table 2.1, the Sun's gravitational energy reservoir must have released altogether a total of $\sim 2.4 \times 10^{41}$ joules of energy during its contraction from a highly dispersed state to the present state. The Sun's present luminosity (that is the rate of output of energy) is 4×10^{26} watts (1 watt = 1 joule per second, a measure of power). Therefore if this luminosity was not significantly different in the past, the Sun cannot have been in existence for longer than

$$\frac{2.4 \times 10^{41}}{4 \times 10^{26}} \text{ seconds} = 6 \times 10^{14} \text{ seconds}$$

$$\cong 20 \text{ million years.}$$

This age is, however, considerably smaller than the age of the Earth estimated at around 4,600 million years from the geological evidence. The Kelvin–Helmholtz hypothesis was therefore considered untenable. As we shall see in Chapter 5, the source of Solar Energy is of a different kind.

The Strength of Gravity

We end this discussion of Newtonian gravity with a few examples to illustrate how strong the force of gravity can be in astronomical circumstances.

Let us begin with our Earth. In the last section we saw that the total work that needs to be done in order to lift one kilogram of matter to an *infinite* height is 6.3×10^7 joules. We can look upon this work in a slightly different way. Suppose the piece of matter is thrown vertically up with a certain speed V. It will rise but at the same time begin to slow down. The reason is that as it rises it begins to acquire gravitational potential energy. This energy has come from the work done by the body against the force of gravity and it must be subtracted from the energy reservoir of the body. Now the only energy reservoir the body started with was of *kinetic energy*. Newtonian dynamics tells us that the kinetic energy of a body of mass m

and velocity V is given by

$$T = \tfrac{1}{2}mV^2.$$

As the body rises this store of kinetic energy is depleted and the speed of the body is reduced.

The interesting question is whether the body will come to rest before it has 'gone to infinity'. This will happen if the kinetic energy reservoir of the body was insufficient to carry it all the way to infinity. To decide whether this energy reservoir is sufficient or not we have to compare the kinetic energy T for 1 kg of matter with the amount of work needed to carry it to infinity. The speed V at which these two are equal is obtained by solving the equation

$$\tfrac{1}{2}V^2 = \frac{GM_\oplus}{R_\oplus}.$$

If we substitute the values of G, $M_\oplus$, and $R_\oplus$ in this formula we get $V = 11.2 \, \mathrm{km \, s^{-1}}$. This speed is known as the *escape speed* for the Earth. In order to escape from the gravitational pull of the Earth the speed of the body cannot be less than the escape speed. An astronaut wanting to go to the Moon or to distances beyond must have enough rocket power to surmount this escape barrier.

The escape speed for any spherical astronomical object is given by a similar formula. It is indicative of how strong the pull of gravity is on the surface of the object. The larger the value of V the bigger is the gravitational pull of the object. In the Table 2.2 below we give the escape speeds for several astronomical objects.

The last entry in the table is of special interest to us. For the black hole, $V = c$, the speed of light. The English astronomer John Mitchell first pointed out in 1783 the possibility that an astronomical object may have such a strong surface gravity that even light cannot escape from it. In 1799

Table 2.2 Escape Speeds for some Astronomical Objects

Astronomical object	Escape speed (in km s^{-1})
Moon	2.4
Earth	11.2
Jupiter	61
Sun	620
Typical white dwarf	5000
Typical neutron star	1.3×10^5
Black hole	c

Laplace also gave a mathematical proof that such objects could in principle exist. To an outside observer these objects will be invisible (hence the name *black* hole), although he can infer their existence from their gravitational properties. We will encounter black holes again in Chapter 4.

From the strength of gravity *outside* the object let us now turn our attention to the strength of gravity *within* the object. Here again we will start with a terrestrial situation. Instead of the astronaut we now consider a diver.

Any person going deep under water is conscious of increasing pressure. At a depth of 34 feet, the pressure has increased to twice its value on the surface of the water where it is the normal atmospheric pressure. The pressure inside the water goes on increasing above this atmospheric value as the diver goes deeper and deeper. Why does the pressure increase?

In Fig. 2.6 we see the reason why. If we consider the column of water up to any given depth we can ask the question as to how it stays put in spite of the Earth's gravity. The Earth's gravity must pull the column of water down, unless there is a counterbalancing force. This force is supplied by the pressure inside the water. Notice that the column is being pressed *down* by the atmospheric pressure above and it is being pushed *up* by the water pressure below. Since this upward pressure must balance not only the downward pressure of the atmosphere but also the downward force of the weight of the column, it must exceed the atmospheric pressure by an amount which depends on the weight of the column supported. So the deeper the diver goes the higher is the pressure encountered by him.

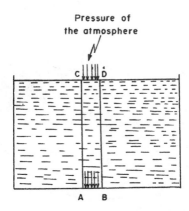

Fig. 2.6. The upward pressure on the bottom surface AB of the water column supports the weight of the column above it as well as the downward pressure of the atmosphere acting at the upper end CD.

If we apply the same ideas to stars like the Sun, we again see how the force of gravity in the interior of the object can demand huge pressures for internal equilibrium. For example, if we consider a concentric spherical layer in the deep interior of the Sun, as shown in Fig. 2.7, we find that the outward pressure on this layer shown by continuous arrows must be able to support the weight of the matter outside this layer. Since this weight grows as we go inwards to the centre of the star the pressure needed for support also grows.

Calculations of the internal structure of the Sun show that to maintain it in equilibrium under its own gravity, the pressure at the centre must be approximately 10^{16} Newtons/(metre)^2, that is, about a hundred thousand million times the normal pressure of the atmosphere on the surface of the Earth.

This enormous value of the pressure required for stellar equilibrium conveys to us some idea of how large the force of gravity in a large astronomical object can be. To see this result in an even more dramatic light, let us imagine what would happen if, by magic, we 'switched' off the internal pressures inside the Sun. In terms of Fig. 2.7, the continuous outward directed arrows disappear and the Sun under its self-gravity begins to contract.

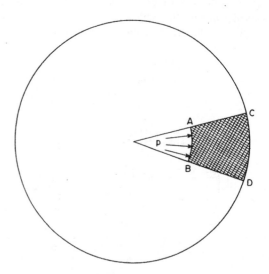

Fig. 2.7. As in Fig. 2.6, the pressure in the interior of the star supports the weight of the outer portions of matter in the star, shown by the shaded region.

The contraction is slow to start with but picks up rapidly because gravity becomes stronger and stronger. Newton's inverse square law tells us that as any two attracting masses come closer their force of gravity on each other increases. The growth of this force within our contracting Sun is catastrophic; for the entire Sun shrinks to a point in a matter of 28 minutes!

This, of course, is only a thought experiment designed to show how important the internal pressures are for the equilibrium of stars. If gravity is allowed to have an upper hand, it becomes more and more powerful as the star contracts and a stage may come when the star's internal pressures can no longer offer an effective counter to its gravity. This leads to catastrophic contraction. This gravity induced implosion is often called *gravitational collapse.*

In Chapter 5 we will study the circumstances which lead to the breakdown of stellar equilibrium. The key to the violent events associated with stars lies in this breakdown.

Conclusion

This completes our survey of Newtonian gravity. We have seen how gravity provides the force which keeps the planets, comets, and satellites of the Solar System in regular motion. Although in our historical approach we emphasized the role of gravity in accurately describing the motion of these objects with clockwork precision, our aim in so doing was to generate confidence in the reader's mind about the overall soundness of the Newtonian picture. For we want to rely on this picture in describing more dramatic events than those of the Solar System, events requiring vast reservoirs of energy. To this end we have also seen how the force of gravity can build up enormous reservoirs of energy in space.

The secret of many energetic or violent phenomena in astronomy is believed to lie in these big gravity reservoirs. The strength of the force of gravity on the surface of or inside a massive object gives us an indication of the important role gravity is expected to play in the equilibrium or otherwise of heavenly bodies.

The considerations of this chapter have been based on Newton's laws of motion and gravitation. In the following chapter we will look at an alternative picture of gravity which is believed to be internally more self-consistent and observationally more accurate than the Newtonian picture. We have already seen, through the example of the motion of the planet Mercury, that all is not well with the Newtonian picture. Nevertheless, this picture is still very useful, and in most astronomical circumstances reasonably accurate. After we have seen the alternative picture presented in the next chapter we will be able to reassure ourselves on this point, for there we will learn exactly when the Newtonian picture is reliable and when it is not.

3 Gravity According to Einstein

> To Einstein, hair and violin,
> we give our final nod;
> though understood by just two folks,
> himself – and sometimes – God!
>
> Jack C. Rossetter (from *The Mathematics Teacher*, November 1950, p. 341)

For nearly two centuries Newtonian gravity enjoyed a continuous run of successes. The important position originally occupied by gravity in overall physics, however, declined during this period. Laboratory experiments took precedence over astronomical observations and the former led to important discoveries which pushed the electromagnetic theory to its place of pride in physics. It was the electromagnetic theory which in the beginning of the present century inspired Albert Einstein to think of the special theory of relativity.

The special theory of relativity, outlined in Appendix B, drastically revised the physicists' concepts of space, time, and motion. Einstein found, for example, that the inherent symmetries present in the laws of motion of Galileo and Newton were inconsistent with the symmetries of the basic laws of electromagnetic theory discovered by James Clerk Maxwell in the 1860s. To make fundamental physics self-consistent it was necessary to modify either the well established laws of dynamics or the more recent laws of electromagnetic theory. Einstein took the bold step and modified the former. As explained in Appendix B the modification of laws of motion demanded a revision of a more basic type – of how physical observers measure spatial displacements and time intervals.

What is Gravity?

Special relativity was proposed in 1905. In the following decade (1905–15) Einstein concerned himself with another fundamental question which again brought him into conflict with Newtonian physics. Einstein wanted to find an answer to the question 'what is gravity?'.

In Chapter 2 we saw that the inverse square law adequately describes the behaviour of the gravitational force between any two chunks of matter.

Why should any two chunks of matter situated well away from each other be subjected to such a force of attraction? Why should this force be describable by the inverse square law? And, what is most important: how does the gravitational force communicate itself *instantaneously* across the vast astronomical distances?

'*Hypotheses non fingo*' (I make no hypotheses) was the reply that Newton himself gave when faced with such questions. Newton's approach to gravity was essentially empirical. As we saw in Chapter 2, the rules according to which planets move round the Sun (the rules known as Kepler's laws) implied the inverse square law for the force of attraction. Why such a law operates is a deeper question which Newton declined to answer or to investigate.

After proposing the special theory of relativity Einstein began to consider these questions. And he soon discovered that the inverse square law of gravitation cannot coexist consistently with the special theory of relativity. Let us look at two aspects of this problem to appreciate the reason why.

One of the major achievements of the special theory of relativity was the demonstration that there exists a fundamental barrier of speed in nature, a barrier which cannot be exceeded by any material particle or by any physical disturbance travelling from one point to another. This limiting speed is attained by light or by any electromagnetic disturbance which propagates as a wave. As shown in Appendix B, this limiting speed, denoted usually by c, plays a crucial role in relating the space-time measurements of *inertial* observers, observers moving under no forces.

The gravitational interaction as postulated by Newton clearly exceeded this light barrier: its effect across vast astronomical distances was supposed to be instantaneous! It was necessary therefore to modify the Newtonian picture sufficiently to make it conform with the light-speed barrier imposed by special relativity. It may be mentioned here that in its earlier versions the electromagnetic interaction was also described by the inverse square law. Coulomb's experiments in 1785 had led him to suggest that like electric charges (or magnetic poles) repel and unlike electric charges (or magnetic poles) attract each other according to the inverse square law. Subsequently, however, this simple law had to be modified when experiments in the nineteenth century with rapidly moving charges began to produce results inconsistent with its predictions.

That the modification needed in the law of gravitation goes deeper than in the case of Coulomb's law was indicated by the second point of conflict between the Newtonian law and the special theory of relativity. Let us examine this second aspect which brings us back to the role of the above mentioned inertial observers in special relativity.

In Newtonian physics space and time measurements have an absolute character. If two events occur at different points of space at different instants of time, the spatial distance and the time interval between these two points and instants will not depend on who makes these measurements. All observers would come up with the same answer. This Newtonian basis of space-time measurements, which seems intuitively acceptable, was challenged by special relativity. Suppose we have two observers moving under no forces. Such observers move with uniform velocities and are the inertial observers referred to earlier. The new rules of space-time measurements given by special relativity showed that the spatial distances or the time intervals between two events *do not* have the same values for all inertial observers. In Appendix B we see how these measurements are related to the relative speed between two inertial observers.

However, the phenomenon of gravity is such that it prevents us from defining the inertial observers in the first place! For gravity is an ever-present force from which no observer is immune. Any two observers on the surface of the Earth are subject to Earth's gravity. This force of gravity may diminish as we move away from the Earth but it will never completely disappear. Nor can we invent any shielding device which will eliminate gravity in any region of space. Shielding devices which remove electrical or magnetic forces are readily available but gravity has defied any attempts to remove it from any part of space. Thus the inertial observer so fundamental to special relativity does not exist because of the ever-present force of gravity.

Einstein gave an ingenious interpretation to this property of gravity. Realizing that gravity is permanently attached to space, he argued that it in fact describes an intrinsic property of space and time, viz. its geometry. We now consider this radical idea in some detail.

The Geometry of Space and Time

The Oxford dictionary describes geometry as the 'science of properties and relations of magnitudes in space'. The first systematic account of geometry was given by Euclid around 300 B.C. It is Euclid's geometry that is taught in schools to this day and it is this geometry that is most commonly used in everyday life such as in the construction of buildings, bridges, tunnels, etc.

The Euclidean geometry – as any other branch of modern mathematics – starts with a few axioms or postulates. These are statements whose truth is taken for granted and the entire subject rests on them just as a building rests on its foundations. If the postulates are changed the subject which is based on them also changes.

It took mathematicians several centuries to realize that Euclid's postulates could be changed and that geometries other than the Euclidean could be formulated in a logically self-consistent manner. The work of Lobatchevsky (1793–1856), Bolyai (1802–1860), Gauss (1777–1855), and Riemann (1826–66) in the last century led to many non-Euclidean geometries. How they differ from the Euclidean geometry can be appreciated with a few examples.

First let us consider what is meant by a 'straight line'. In Fig. 3.1(a) we see a line (drawn on a plane paper) which is not straight. At each point of the curve, if we draw the tangent to it, the direction of the tangent changes as the point moves along it. In the case of a straight line this direction does not change. In Fig. 3.1(b) we see another way of deciding which curve is straight. Of the lines connecting the two points A and B only the dotted line is straight – it being the line of *shortest length* between A and B. If a rubber band is stretched between A and B it will tend to assume the shortest length and will lie along the dotted line.

To us, accustomed to drawing lines on plane paper, these properties of straight lines are intuitively acceptable. We can also accept Euclid's parallel postulate which tells us that given a straight line *l* and a point P

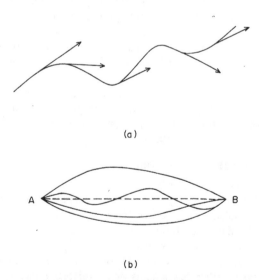

(a)

(b)

Fig. 3.1. In (a) we see a curve with tangents at typical points on it shown by arrows. The tangents point in different directions indicating that the curve is not straight as per the criterion of Euclidean geometry on the plane paper. In (b) we have several lines joining points A and B. The length of the dotted line is the least, as would be indicated by stretching a rubber band along it. The dotted line is straight.

outside it, we can draw *one* and *only one* straight line through P parallel to *l*.

Yet it is not necessary to retain this postulate for a self-consistent geometry. For instance, we can construct a geometry on the assumption that through P *no* line can be drawn parallel to *l*. We can also go the other way: we can assume that through P more than one line can be drawn parallel to *l*. These alternatives become acceptable if we discard our intuitive notion of *flat space* such as the two dimensional space of the plane paper described before.

Imagine, instead, the curved two dimensional surface of a sphere. What would the geometry be like if the lines are constrained to be on the sphere? As shown in Fig. 3.2, we can draw a 'straight line' between *any* two points A and B on the sphere by stretching a rubber band between them. This line is in fact the arc of the great circle passing through A and B. (A great circle is the circle cut out on the spherical surface by a plane passing through the centre of the sphere.)

Now any two great circles intersect and so all straight lines on the spherical surface intersect. This result brings us into conflict with a concept

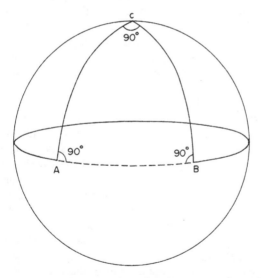

Fig. 3.2. The 'straight line' on the surface of a sphere, connecting typical points A and B on it, is the arc of a great circle joining A and B. The distance between A and B is measured along that arc (shown by dotted line) whose length is *less* than πR, where R is the radius of the sphere. Any two great circles on the sphere intersect so that there are *no* parallel lines in the geometry on the surface of the sphere. The triangle ABC shown here has each of its three angles equal to a right angle.

familiar to us from Euclid's geometry, the concept of parallel lines. Two straight lines drawn on a plane surface are considered parallel if they never meet, even if extended indefinitely in either direction. Clearly this concept does not hold on the surface of a sphere. In other words, *there are no parallel lines* and we have here an example of the first type of violation of Euclid's parallel postulate. Euclid's proofs which make use of parallel lines naturally fall apart in the geometry of the spherical surface. For example, any triangle ABC will have its three angles adding up to a sum *greater* than 180°. The triangle shown in Fig. 3.2 has $\hat{A} + \hat{B} + \hat{C} = 270°$.

Curved surfaces of this type are called surfaces of *positive* curvature. If we had instead elected to modify the parallel postulate in the second of the two ways described before, we would have arrived at a geometry which applies to surfaces of negative curvature. A saddle, or the surface of a jar near its lip, are examples of surfaces of negative curvature. For a triangle ABC on such a surface we have $\hat{A} + \hat{B} + \hat{C}$ less than 180°.

What has all this to do with gravity? The concepts of flat and curved space can be extended to spaces of higher dimensions. For example, the geometry of the three dimensions of space and one dimension of time in which Einstein's special theory of relativity applies, is the geometry of *flat* space. Because of the ever-present gravity this flat space geometry is, according to Einstein, an unrealizable idealization. The geometry of space and time must be of a curved non-Euclidean type. This important conclusion of Einstein is often stated in the form: 'space-time is curved'.

We now examine the physical implications of this statement.*

Motion in a Curved Space-time

In Fig. 3.3 we see the path followed in space by a projectile, say a cannon ball, fired at an angle to the vertical direction. The path shown by the continuous line is a *parabola*. The cannon ball rises to a certain maximum height and then falls to the ground, travelling all the while in the horizontal direction also. How do we account for this path in Newtonian dynamics?

Had there been no gravity, the ball would have travelled with uniform speed along the dotted straight line shown in Fig. 3.3. This is what Newton's first law of motion tells us. Because of Earth's gravity the projectile is pulled towards the Earth and its actual trajectory falls *below* the dotted straight line. How would Einstein regard this situation?

According to Einstein, the dotted straight line has no real status at all since the 'no gravity' situation is never realized in nature. Instead, we argue

---

* There are many types of non-Euclidean geometries. Einstein opted for the geometry of Riemann. The Riemannian geometry has one simplifying property, that the two Euclidean properties of a straight line mentioned before, viz. 'a line of unchanging direction' and 'a line of shortest distance', hold in this geometry also.

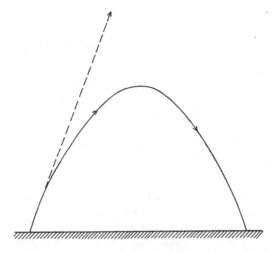

Fig. 3.3. The path of a projectile on the surface of the Earth is a parabola. Had there been no gravity the projectile would have gone along the dotted straight line. (This is a space diagram, not a space-time diagram. The time coordinate has not been shown.)

that the space-time in the neighbourhood of the Earth is curved because of Earth's gravity. In this curved space-time the cannon ball moves *under no forces*. The continuous line – the parabola – is the track along which the particle moves under no forces, that is, in a straight line with uniform speed as required by Newton's first law of motion, provided all measurements are made according to the geometry of curved space-time.

The reader may wonder where *time* has explicitly entered in the above example of the projectile. The tracks we have shown in Fig. 3.3 are tracks in space only, which tell us the successive positions occupied by the projectile in space. If we had drawn a space-*time* diagram, we could have shown in it the location of the projectile at any given time of its flight. A line showing the location of a particle at different times in the space-time diagram is called the *world line* of the particle. Einstein's above rule of 'straightness' for motion under no forces applies to the world lines of moving particles. Thus in the above example, if we had plotted the world line of the projectile moving in the parabolic trajectory of Fig. 3.3, we would have found that it satisfies the criterion of straightness in the curved space-time near the Earth.

Notice first that by treating the space-time as curved Einstein eliminated gravity as a physical force. So we have the following change-over from the Newtonian picture:

NEWTON: Projectile moving in *flat* space-time under the Earth's *force of gravity.*

EINSTEIN: Particle moving in *curved* space-time under *no forces.*

The second point to notice in the above example is the fact that if we change the rules of geometry, the criterion determining whether a curve is straight or not is also modified. Therefore, according to the changed rules of the curved space-time near the Earth, the motion along the continuous curve does describe uniform motion in a straight line.

The sceptic may perform the following experiment to test this idea. On a globe stretch a rubber band between London and Los Angeles. This gives the shortest route between these two cities. Now repeat this procedure on a flat map of the world. The route will be different. No flat map can faithfully reproduce the geometry on the surface of the Earth and hence the latter route is *not* the shortest route between the two cities.

The above example illustrates how different rules of measurement in space can lead to different answers to the question: 'what is a straight line?' Similarly, different rules of measurement in space *and time* lead to different answers to the same question about lines drawn in space-time. The technical name given to the world line of a particle moving under no forces in curved space-time is *'geodesic'*. Geodesics are the straight lines of the curved space-time.

In his *general theory of relativity* Einstein gave a set of equations which relate the geometrical properties of space-time to the distribution of gravitating matter within it. Thus in the above example, Einstein's equations determine the exact nature of space-time geometry near the Earth. The parameters of this geometry are determined by the mass and size of the Earth. Once we determine this geometry we can find out how particles move in it by calculating the appropriate geodesics in it.

The general theory of relativity goes far beyond the special theory (described in Appendix B). Special relativity brought into physics the important notion that space and time together form a joint entity. The measurements of spatial distances and time intervals in the special theory are performed according to flat space geometry. The notion of curvature of space-time and its relation to gravity is the remarkable new feature of the general theory.

In 1916, a few months after Einstein's equations were published, Karl Schwarzschild solved them in order to calculate the nature of curved space-time outside the Sun. Knowing the geometry of this space-time it is then a relatively simple matter to calculate the geodesics in it. *These are the world lines of planets.* A typical planetary world line is shown in Fig. 3.4.

In the space-time diagram given here, the world line spirals up as time

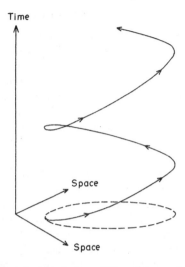

Fig. 3.4. The world line of a typical planet. This spiral curve has a projection in the space dimensions which gives the nearly elliptical orbit of the planet.

passes. Its projection in space gives an almost *elliptical* shape, showing the similarity between the predictions of Newtonian gravity and Einstein's general relativity. The adjective 'almost' needs further elucidation. The ellipse along which the planet moves according to Einstein's theory slowly rotates so that the perihelion advances steadily. This effect is negligible for most planets, being noticeable only in the case of the planet Mercury because its orbit is closest to the Sun and is highly eccentric. General relativity predicts that the perihelion of Mercury advances at an angular rate of 43″ per century.

We now recall from Chapter 2 that this is precisely the effect which had remained unexplained in Newton's gravity. The planetary orbits calculated according to general relativity using the Schwarzschild solution differ from the orbits calculated by the Newtonian theory by small amounts. The biggest estimated difference between the two theories is for the planet Mercury and the observations favour general relativity. Indeed, the circumstance that general relativity accounted almost exactly for this outstanding discrepancy went a long way towards establishing the credibility of this theory.

The smallness of the effect, however, illustrates the fact that the differences between the predictions of the two approaches to gravity are expected to be small within our planetary system. We now discuss the other experimental tests devised to detect such differences.

The Éxperimental Tests of General Relativity

Apart from the precession of the perihelion of Mercury, two other tests were proposed and carried out soon after Einstein put forward the general theory of relativity. In those early days relativity, with its radically different approach to gravity, was not fully appreciated or understood by physicists in general. Among Einstein's contemporaries, the British astronomer Sir Arthur Eddington played the leading role in popularizing relativity and making known its observable consequences. It was Eddington's efforts that prompted astronomers to conduct observations in respect of the above two tests.

One test involves the *bending of light* by a massive object. Just as Fig. 3.3 shows that gravity affects the path of the cannon ball, it should also affect the path of light. In Newtonian gravity, in its original form, the inverse square law was not supposed to apply to light. Indeed, light was believed to be immune to the effects of gravity. However, with the modern inputs from quantum theory, there are reasons for believing that light should be attracted by matter. In fact, we may estimate the effect of Newtonian gravity on light in the following way. Quantum theory tells us that a light wave of frequency v is in fact made of tiny packets of energy, called *photons*, each photon having energy

$$E = hv.$$

According to special relativity (see Appendix B) this energy is equivalent to a mass m, given by the relation

$$E = mc^2.$$

In other words, a region of space containing a given quantity of energy possesses an equivalent mass obtained by dividing the amount of energy by the square of the speed of light. By this rule, for example, ninety million million joules of energy are equivalent to about one gram of mass. We therefore associate a mass

$$m = \frac{hv}{c^2}$$

with a photon of frequency v.

Figure 3.5 shows how such a stream of photons skirting the surface of a massive object would be bent by its gravity. The effect for the Sun is very very small, the bending angle being given by the formula

$$\alpha = \frac{2GM_\odot}{c^2 R_\odot}.$$

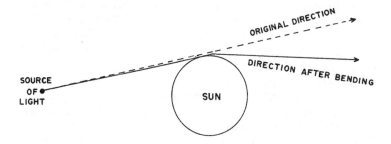

Fig. 3.5. The ray of light is bent by the gravitational force of a massive object.

This bending angle is shown grossly exaggerated in Fig. 3.5. When we substitute the values of the mass $M_\odot$ and the radius $R_\odot$ of the Sun, we get $\alpha = 0.875''$ (arc second).

General relativity looks at this effect differently. Light travels in straight lines even in the presence of gravity. Of course, gravity alters the space-time geometry and makes it non-Euclidean and so the light tracks are also different in the neighbourhood of a massive object. The world line of light follows what is called a *null geodesic*.* To find how light would travel as it grazes the solar limb we have to find appropriate null geodesic in the space-time geometry of Schwarzschild. Qualitatively the effect is similar to that of Fig. 3.5. The net bending angle is, however, *double* the Newtonian value and equals

$$\alpha = \frac{4GM}{c^2 R} = 1.75''.$$

What does this result mean in the practical language of the astronomer? Consider a star being observed by an astronomer on the Earth, and suppose at some stage of the observation the line of sight to the star crosses the solar limb. The bending effect described above is at its maximum when the line of sight to the star grazes the surface of the Sun, and because of the bending the star image shifts. The bending angle α denotes the shift in the apparent position of the star. But how can the astronomer observe a star so close in its direction to the Sun?

The bending can be observed at a *total solar eclipse* when the star is occulted by the Sun. Realizing the implications of such an experiment Eddington prompted the Astronomer Royal Sir Frank Dyson to provide

* The adjective 'null' comes from the fact that according to the rules of special relativity, explained in Appendix B, the distance between any two points of space-time connected by a light ray is zero. The same rule continues to hold in the curved space-time of general relativity.

support for the expeditions to observe the total solar eclipse on 29 May 1919 at Sobral in Brazil and at Principe, a small island in the Gulf of Guinea. Although Eddington's results did seem to support the predictions of general relativity the full range of uncertainty of the observations was not appreciated in those days.

It is now known that the refractive effects near the surface of the Sun can also produce comparable bending effects on rays of visual light. These effects are negligible, however, on radio or microwaves. The techniques of very long base line interferometry have now made it possible to measure α very accurately at these longer wavelengths, the observational uncertainty being of the order of 1 per cent. In 1975–6 experiments were made to measure α for the quasar 3C 279 as it was occulted by the Sun. The results support the predictions of general relativity. (Since the radiation from the Sun in radio or microwaves is small, it is not necessary for these experiments to be conducted at a total solar eclipse.)

The second test brings us the possible effect of non-Euclidean geometry on the measurements of time. Suppose we have two observers A and B at rest relative to each other, communicating with each other with light waves of frequency v. Of these A is located far away from a gravitating object of mass M and radius R, while B is located on its surface. Now, the effects of non-Euclidean geometry are expected to be large at B (near the object) and small at A (away from the object). These effects show up in the following way.

Since B is sending waves of frequency v, he will send v waves per second as measured by his own watch. In other words, two successive wavecrests will leave B at an interval of $1/v$ second. What is the time gap measured by A as he receives these same wavecrests? Calculations of general relativity show that the time gap is *increased* by a fraction z of the original, where z is given in terms of the mass and radius of the gravitating object by

$$z = \frac{1}{\sqrt{1 - \dfrac{2GM}{c^2 R}}} - 1.$$

That is, the gap between the arrival times of successive wavecrests at A is increased to

$$\frac{1}{v} \times (1 + z),$$

and the frequency measured by A is reduced to

$$\frac{v}{1 + z}.$$

This reduction in frequency corresponds to an *increase* in the wavelength of the wave as measured by A. This is because the frequency of a wave multiplied by its wavelength tells us how far the wave advances in time, that is, the speed of the wave. In this case the speeds of the wave as measured by both A and B are the same. Therefore, compared to the wavelength sent out by B, the wavelength received by A is increased by the factor $(1 + z)$.

The quantity z is called the *red-shift*. If instead of sending waves of one frequency, B were sending a range of frequencies in the visible part of the spectrum, A would notice a shift in this spectrum towards the *red* end. Since the cause of the red-shift is gravity, the effect is known as the *gravitational red-shift*.

In the above example light waves leave from a region of strong gravity and go to the region of weak gravity. What would happen in the reverse situation, in the case where the source of the waves is located in a region of weak gravity and the receiver is in a region of strong gravity? Suppose in the above example we have waves sent out by A and received by B. The calculations of general relativity then show that by the time such waves reach A their frequency is *increased* by the factor $(1 + z)$ given above, and their wavelength is reduced by the same factor. Therefore, if A is sending out to B a spectrum of visible light, such a spectrum is shifted towards the blue end when received by B. This phenomenon is therefore called *gravitational blue-shift*.

By astronomical standards, we on the surface of the Earth are located in regions of weak gravity compared to that on a typical astronomical source like a star. We are therefore more likely to see gravitational red-shifts than blue-shifts in the spectra of astronomical objects. In this sense we are more like the observer A than B.

Such spectral shifts should be seen whenever light travels in a region of inhomogeneous gravity. When z is very small compared to one, say less than 1 per cent, we can approximate its value by

$$z = \frac{GM}{c^2 R}.$$

Thus if light from a white dwarf is examined for spectral lines we should notice *red-shifts* of the order of 10^{-4} to 10^{-3}. Such red-shifts have now been confirmed, for example, for the white dwarfs Sirius B $(z \sim 3 \times 10^{-4})$ and 40 Eridani B $(z \sim 7 \times 10^{-5})$.

This confirmation, however, *does not* unequivocally single out general relativity in preference to Newtonian gravity. If we consider our modified version of Newtonian gravity wherein light is supposed to be made of photons, we arrive at the same answer in the cases where z is small compared to one. In Box 3.1 we show a simple derivation of this result.

Box 3.1 Gravitational red-shift in Newtonian gravity

By a slight modification of Newtonian gravity it is possible to account for the phenomenon of gravitational red-shift. The modification consists in recognizing that light of frequency v is made of packets of energy hv, where h is known as Planck's constant. The typical packet, called the *photon*, has therefore an equivalent mass of

$$m = \frac{hv}{c^2}.$$

This mass makes the photon subject to gravity just as any ordinary particle is, although Newton himself did not assume that light would be subject to the law of gravitation.

Imagine a photon leaving the surface of a massive object of mass M and radius R and escaping to infinity. We have already seen in Box 2.3 that this entails work to be done on the part of the photon to the tune of GM/R per unit mass, i.e. a total work of GMm/R. This work is done at the expense of the photon's energy. This loss of energy of the photon results in a reduction of its frequency to v', so that

$$hv - hv' = \frac{GMm}{R}$$

$$= \frac{GMhv}{Rc^2}.$$

Hence the change in frequency is given by

$$\Delta v = v - v' = \frac{GM}{Rc^2} v,$$

corresponding to a fractional decrease in frequency by the amount GM/Rc^2.

Recent improvements in observing techniques have added two more to the list of experimental tests of general relativity in the Solar System which make use of the geometrical properties of the Schwarzschild space-time. One test is concerned with the relativistic prediction that if a radar signal is bounced off a planet while it is at superior conjunction with respect to the Sun, the signal grazes the limb of the Sun and it takes a slightly longer time to complete the two-way journey than it would otherwise do. The expected delays from Mercury are in the region of 200–250 μs (1 μs equals a *millionth* part of a second). Such delays have been confirmed, not for the Earth–Mercury round trip, but for signals bounced from the Earth by the

Mariner 6 and 7 spacecrafts in superior conjunction. The estimated delay of 200 μs has been observed within 3 per cent experimental errors.

The second test makes use of the knowledge of the accurate distance between the Earth and the Moon by laser ranging. Such high accuracy is achieved in these experiments that it is possible to tell this distance within ~ 15 cm, with the help of ruby lasers reflected from the Moon. This in turn enables us to check whether the motion of the Moon is subject strictly to the laws of relativity or whether small additional effects not predicted by relativity are present. Here again, the predicted positions of relativity are accurately verified.

There is another possible test of general relativity which we will discuss in the next chapter since it brings in the effects of rotation which are not covered by Schwarzschild's solution. We will also postpone the discussion of the phenomenon of another relativistic effect, viz. gravitational radiation, to an appropriate later stage.

The Choice between Newton and Einstein

The effects discussed in this chapter show how minute is the difference between the predictions of general relativity and Newtonian gravity. The general rule of thumb is that the gravitational effects at a distance R from a gravitating mass M are considered small if the parameter

$$\alpha = \frac{GM}{c^2 R}$$

is small compared to unity – say less than 1 per cent – and in the cases of such small effects the predictions of the two theories differ by very small amounts. In such cases the simpler Newtonian framework is preferable to relativity from a practical point of view.

For example, in the Solar System α has the largest value on the surface of the Sun, where $\alpha = 2 \times 10^{-6}$. This small value of α suggests that we rely on the use of Newtonian gravity within the Solar System. The highly accurate manoeuvres of the lunar and interplanetary spacecrafts justifies this confidence in Newtonian gravity.

Nevertheless we must not lose sight of the conceptual shortcomings of Newtonian gravity which led Einstein to think of general relativity. In particular, whenever we encounter situations of strong gravitational effects (where α is close to unity, say lying between 0.1 and 1) we should begin to suspect calculations based on Newtonian gravity and switch over to general relativity. Such cases are likely to arise in astronomy and to deal with these we will consider in the next chapter the situations where the gravitational effects are strong.

4 The Physics of Black Holes

I'm very well acquainted too with matters mathematical,
I understand equations, both ordinary and differential,
About Riemann's geometry I'm teeming with a lot o'news –
With many new facts about the square of the hypotenuse.

I'm very good at black holes, be they charged or spinning,
About singularities I know more than Penrose and Hawking;
In short, in matters vegetable, animal and mineral,
I am the very model of a modern Major General.

From the song of the Major General in 'The Pirates of Penzance'
– adapted from 1880 to 1980

In the last chapter we saw that when gravitational effects are weak, the predictions of Newtonian laws of gravity and of general relativity almost agree. There are in fact differences in their predictions of an order characterized by the dimensionless parameter $\alpha = GM/c^2R$, which is small for weak gravity. The observational tests in the Solar System are designed to measure quantities of this order so that they may distinguish between the two ways of describing gravity.

While discussing Newtonian gravity in Chapter 2, we came across this parameter in a different way. There we found that the concept of escape velocity V helps us in measuring the strength of gravity on the surface of any massive object. The larger the value of V, the more difficult it is for a projectile (or a spaceship) to leave the surface of the object and escape from its gravitational pull altogether. The condition that V becomes equal to c, the speed of light, tells us again that the parameter α is no longer very small compared to 1:

$$\alpha = \frac{GM}{c^2R} = \frac{1}{2}.$$

It is therefore of interest to ask the question as to whether objects with a high value of this parameter exist in nature – objects with a ratio of mass (M) to radius (R) considerably greater than that for the Earth or the Sun. The ratio M/R can be increased in two ways, either by lowering R or by increasing M. It is tempting to think that the former condition could lead us to cases of strong gravity in atomic or sub-atomic (nuclear) physics. Certainly, the radii of the so-called elementary particles are believed to lie in the region of 10^{-16}–10^{-14} m. However, the masses of these particles are so small that the value of GM/c^2R is in the region of 10^{-40}. This small value indicates the weakness of gravity at the atomic and sub-atomic level.

The other alternative, of large M, takes us to astronomy. Here, simply by raising M we may not hope to achieve a large value of M/R, because a massive object may also be very large in size. There is, however, a property of gravity which comes into play at large M that makes it possible to achieve situations of strong gravity. We will now consider this property in some detail.

Gravitational Collapse

In Chapter 2, while discussing the equilibrium of stars, we emphasized the importance of internal pressures in providing sufficient opposition to the contracting tendency of a star's self-gravity. Withdraw these pressures from the interior of the Sun and it would collapse in a matter of minutes.

Now, it is possible to visualize situations in the Universe where the internal pressures (if not withdrawn altogether) are totally inadequate to maintàin the equilibrium of a massive body. A crude way of seeing why this happens is with the help of the concept of energy. In Chapter 2 we saw that the gravitational energy of a body of mass M increases as M^2. The nuclear energy (or any other source of energy) within the body, on the other hand, is proportional to the mass M. Therefore as we go on increasing the value of M, the gravitational energy rises at a more rapid rate than does the nuclear energy. Hence in any scenario where equilibrium between these two opposing forces is concerned, detailed calculations show up a limiting mass M_c with the following property. For $M < M_c$ it is possible for the gravity opposing forces to generate strong enough pressures to maintain the body in equilibrium; but for $M > M_c$ this is impossible. The actual value of the critical mass M_c depends on what type of opposing force is considered relevant within the object.

For example, in the 1930s Chandrasekhar showed that the critical mass for white dwarf stars is $1.44\,M_\odot$. Inside a white dwarf the electrons provide a special kind of pressure known as *degenerate pressure* which arises from their being closely packed together. The density of a white dwarf can be as high as a million times that of water. In recent years astrophysicists have considered matter in an even denser form – with densities a thousand million times those in a white dwarf. This matter exists in the form of degenerate neutrons. The limiting mass M_c in this case lies in the range $2\,M_\odot - 3\,M_\odot$.

These examples emphasize the limitations of forces opposing gravity. If very massive objects do exist in nature, with masses exceeding any M_c imaginable on physical grounds, then such objects must inevitably shrink and we are led to the gravitational collapse described briefly in Chapter 2. We now investigate this phenomenon in more detail, within the framework of general relativity.

We will consider a spherical body of mass M and radius R. One result of relativity which we encountered in the previous chapter becomes more and more noticeable as such a body contracts. This is the effect of gravitational red-shift wherein the light from the surface of the body undergoes a fractional increase in wavelength by the time it reaches the remote observer. This fractional increase z (called the red-shift) is given in terms of the mass and the radius of the body by the formula

$$z = \frac{1}{\sqrt{1 - \dfrac{2GM}{c^2 R}}} - 1.$$

This formula tells us that as the body contracts slowly R decreases and therefore z increases. The light emitted from the contracting body will therefore appear redder and redder to the remote observer.

There is another effect which this remote observer will begin to notice about the contracting body: its growing faintness. Recalling the result mentioned in the last chapter (see page 46) that the energy of a light photon is in direct proportion to its frequency, we see that every photon leaving the surface of the body is *degraded* in energy by the factor $1/(1+z)$. For $z = 1$, each photon will arrive at the remote observer with only *half* its original energy. Naturally the object would appear faint to this observer.

The actual effect is further enhanced when we take into account the time dilatation caused by gravitation. Recall again from Chapter 3 that the gravitational red-shift effect arose in the first place because the intervals of emission of light wavecrests from the surface were not equal to the intervals of reception of these same wavecrests: the latter were longer by the factor $(1 + z)$. Therefore if the body emits a certain number of photons, say 100 photons, over a period of one second as measured by an observer on its surface, these 100 photons will arrive at the remote observer over the longer period of $(1 + z)$ seconds. The *rate* at which the energy from the body is emitted is correspondingly reduced by the factor $1/(1+z)$, as estimated by this observer.

The two effects taken together cause a reduction in the observed flux of light of the body by the factor

$$f = \frac{1}{(1+z)^2}.$$

For $z = 1, f = 25$ per cent.

These effects are based on the assumption that the body is contracting slowly. This assumption is correct in the early stages of gravitational collapse – but not in the later stages. In the later stages the body begins to contract *rapidly* so that with respect to the remote observer its surface is *not* static; rather it is moving away from him. In Appendix C we have discussed what happens to light from a source of light receding from the observer. The effect of such a relative motion of speed V is to increase the wavelength of light received by the observer by the factor

$$1 + z = \sqrt{\frac{1 + \dfrac{V}{c}}{1 - \dfrac{V}{c}}}.$$

Known as the *Doppler effect*, this result also must be taken into account in our discussion of the collapsing object. The above result of Appendix C was derived in the space-time of special relativity. The derivation of the corresponding result in general relativity is more complicated because the Doppler red-shift and the gravitational red-shift get inextricably mixed.

The outcome of such a calculation, however, is to *increase* the total red-shift over that given by the formula for gravitational red-shift alone (see p. 54). In Fig. 4.1 we plot the fractional drop in the flux of radiation l and the increase in the red-shift z of the collapsing object, against the time kept by the remote observer.

For our understanding we also plot the fractional drop in the radial size R of the collapsing object* as measured by the remote observer, although in practice this may be a much more difficult proposition than measuring l and z. Notice that z increases while R and l decrease, as expected from our discussion above. If the time axis were continued to *infinity*, l would fall to zero and z would rise to infinity. What would happen to R?

The value of R decreases, but not to zero! It goes down to

$$R_S = \frac{2GM}{c^2}$$

---

* In a non-Euclidean geometry the meaning of the radial size is not the same as in Euclidean geometry. Here by R is meant the size which gives the total surface area of the spherical object as $4\pi R^2$. In a non-Euclidean geometry R may not equal the actual measured radius R' (equalling the distance from the centre to the surface) of the sphere. In other words, the surface area of a sphere of radius R' is not $4\pi R'^2$ but it is $4\pi R^2$.

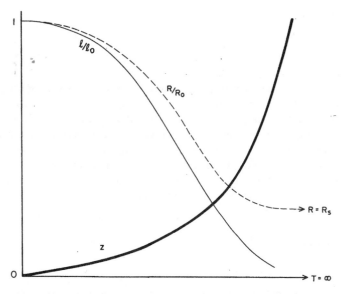

Fig. 4.1.   A schematic plot showing the drop in the light flux (l/l_0) of a collapsing object (continuous line) and the rise in its red-shift (thick line) as observed by a remote observer at different times. The dotted line shows that as noted by this observer the radius of the object never approaches the black hole radius R_s throughout his lifetime, howsoever long it may be. Here l_0 and R_0 denote the starting values of l and R.

as the time kept by the remote observer goes to infinity. This is to be expected, at any rate on the basis of the red-shift formula on page 54. There as R tends to R_s, z rises to infinity.

This means that the external observer will notice the body becoming fainter and fainter, and its spectrum getting redder and redder, as R approaches R_s. The actual value of R_s is *never* achieved in the observer's lifetime no matter how long he lives. The object will, however, disappear from his detecting instruments at some finite stage of his existence as its flux drops below the threshold for detection: to all intents and purposes the object becomes *black*. This is the circumstance which, from a practical point of view, may be identified with the formation of a *black hole*.

From the strict mathematical point of view of the relativist, however, our spherical body is said to become a black hole only when its radial size *equals* R_s. At $R = R_s$, we notice from Fig. 4.1 that the body has zero flux ($l = 0$) and infinite red-shift ($z = \infty$). Thus the body is *totally black*. Any light signal emitted by the body at $R = R_s$ or at subsequent moments during the collapse is unable to leave the boundary $R = R_s$. This boundary is called the Schwarzschild barrier to indicate that it was originally the solution of Karl Schwarzschild that brought out the significance of R_s. Nevertheless, as we shall shortly see, black holes may be more general than the *Schwarzschild black hole* described here.

The lack of communication with the outside world (through light or other material signals) is a property common to all black holes. This property is known as the appearance of an *event horizon*. Just as the curved surface of the Earth leads to the existence of a horizon which limits the range of vision of a navigator on the high seas, so the curved geometry of the black hole leads to the formation of an event horizon which hides from the external observer the part of space-time in its interior.

It is worth emphasizing two points of caution. First, we note that so far as 'outside observers' like us are concerned *no* astronomical body is ever going to become a black hole in the strict mathematical sense described above. As Fig. 4.1 shows, the signals sent at $R = R_s$ are not going to reach us ever. So, any astronomical techniques commonly used for studying the heavenly bodies only tell us about the object *prior* to its reaching $R = R_s$. Any assertion like 'the star has become a black hole' or 'there exists a massive black hole in a galaxy' therefore rests on surmise rather than direct observation of fact. The nearest that one can come to the above statements is from a *practical* rather than the *exact* point of view described above. Figure 4.1 tells us that for all practical purposes the object does become black as R approaches R_s. In the case of a spherical ball of dust (pressure free matter) it can be shown that the light flux from the object drops by a significant fraction, say to a tenth of its value, in a time-scale of the order

$$\tau = \frac{2GM}{c^3} = \frac{R_s}{c}.$$

Thus for a body of mass $M_\odot$ (the mass of the Sun) this time-scale is as small as a few millionths of a second. Such a body will therefore rapidly become very very faint as its surface approaches the Schwarzschild barrier. This difference between the *de facto* and the *de jure* status of black holes is often slurred over in astronomical literature.

The second point to notice here is the superficial similarity between a Newtonian black hole and a relativistic black hole. If we adopt the point of view of Mitchell or Laplace (cf. Chapter 2) we arrive at the value $R = R_s$ for the radius of a body whose surface escape speed equals c, the speed of light. Thus light from the surface of a body with $R < R_s$ cannot go far away from the body. Hence to a remote observer the body looks black. This Newtonian interpretation of a black hole is *mild* compared to its relativistic counterpart. For, even though $R < R_s$ for such a body, the Newtonian interpretation tells us that light *can* leave its surface. A light ray from the surface of such a body will go some distance away from it before falling back on it, just as a stone thrown up from the surface of the Earth eventually falls back on it. Similarly an observer sufficiently close to the object, but outside it, can detect its existence by light rays. This is not as a few millionths of a second. Such a body will therefore rapidly become impenetrable barrier. Light cannot leave this barrier even to go out for a brief sortie – it cannot leave the barrier *at all*. An observer even located near the black hole cannot 'see' it as he might a Newtonian black hole.

As mentioned in Chapter 3, general relativity is conceptually more self-consistent and in better agreement with observations than Newtonian gravity. In the case of strong gravitational effects, therefore, we will place our faith in relativity and our subsequent discussion of black holes in this chapter will be within the framework of general relativity.

A Black Hole has No Hair!

The collapsing body in our discussion above was of a highly special nature: we had assumed the body to be spherical. The problem of spherical collapse can be solved exactly within the framework of Einstein's theory. Thus to the outside observer the geometry of space-time is given by Schwarzschild's solution and this solution is completely characterized by the mass M of the object. By firing test particles at the black hole, the external observer can, in principle, measure the mass of the black hole even though he cannot see it.

However, it is too much to expect nature to be so obliging as to provide us with exactly spherical massive objects. A typical massive object may have many built-in irregularities. For instance, it may be rotating, it may have electric currents and magnetic fields in it, its shape may be non-spherical, even jagged. Some or all of these features are present in astronomical systems. It is therefore pertinent to ask about the fate of such irregularities as the object collapses. Does the object become a highly irregularly shaped black hole?

Here, unfortunately, general relativity has not yet provided the answer. Einstein's equations are so complicated that present computer technology is nowhere near the level at which it could claim to be able to solve the problem of collapse of a general – odd shaped – object. Analytical methods also are not capable of delivering the answer. That its strong self-gravity will make the object shrink is beyond doubt. But whether the object will continue to maintain a coherent shape is not so clear. If it falls apart the problem of collapse altogether changes in character.

This lack of exact solution has not prevented theoreticians from making conjectures about the ultimate fate of such collapsing objects provided they hold together. These conjectures are based on the detailed investigations of a considerably more limited state of affairs described below. Suppose instead of having a perfect sphere, we have slight departure. from spherical symmetry. Thus we can imagine the small irregularities to be just like the hills and valleys on the surface of the Earth. We can also include rotation and the electromagnetic fields within the object. All these (and any other imaginable *small* disturbances) are assumed to be so small that they do not produce any significant effect on the overall geometry of space-time outside the object. Under these assumptions, it is possible to *linearize* Einstein's equations – a procedure which enables mathematicians to obtain their solution exactly. R. H. Price who considered this problem in the late 1960s arrived at the following remarkable conclusion.

Each initial irregularity in the collapsing object tends to be smoothed out, with the exception of a select few, as the object collapses. The smoothing out process takes place through radiation and the critical property which determines which information survives and what is radiated away is the *spin* of the basic interaction conveying the information. Price's theorem relates the spin to the 'moments' which mathematically measure the departures from spherical symmetry of the various irregularities. These moments tell the outside observer what type of irregularities are present in the object. *For survival, the order of the moment must be less than the spin.*

The notion of 'spin' of an interaction became established in physics with the advent of quantum theory. According to current ideas the physical effects of an interaction are conveyed from one point to another by small particles. The properties of the interactions are expressed through certain specific characteristics of these particles like their mass, electric charge, etc. Spin is one such characteristic and it is usually expressed in the atomic unit of $\hbar = h/2\pi$, where h is Planck's constant.

For example, the electromagnetic interaction is conveyed through the particle of light, the photon. The photon has spin 1. Price's theorem therefore tells us that all electrical and magnetic moments of the system of order 1 and over are radiated away, leaving only the information about the moment zero, which is conveyed by the electric charge Q. Gravity itself is transmitted through the *graviton*, a particle of spin 2. According to Price, the surviving moments during collapse are of order 0 and 1. The moment of order zero is the mass M of the object while the moment of order 1 is its angular momentum S.

We have therefore the following picture of the gravitational collapse of a nearly spherical body. Imagine the body to be in communication with the outside observer through the various physical interactions. These interactions convey to the observer the details of the body structure. These details, however, begin to get blurred as the collapse proceeds, until close to the black hole stage only those details remain which are communicable to the outside observer as moments of order *less* than the spins of the transmitting interactions.

The moral of Price's theorem is that during the collapse of the object most of its irregularities are lost. What remains noticeable to the outside observer, as the body becomes a black hole, is the set of a few hard-core properties only. If we limit our attention to gravity and electromagnetic theory only, then the set of surviving properties consists simply of the mass M, the charge Q, and the angular momentum S of the black hole. Of the four basic interactions known to the physicists only these two (gravity and electromagnetic) interactions are of long range and are likely to be of importance in conveying information about the black hole to the remote observer.

If the same conclusions hold in the collapse of the most general character – and it is a big 'if' – then we arrive at a picture of the most general black hole about which very little detailed structural information is available to the remote observer. This state of affairs is expressed by John Wheeler in these oft-quoted words: 'A black hole has no hair!'

The Kerr–Newman Black Hole

Soon after Schwarzschild obtained the space-time geometry outside a spherical object of mass M, two scientists, H. Reissner in 1916 and G. Nordström in 1918, independently solved Einstein's equations to find the space-time geometry outside a spherical object of mass M and charge Q. Then in 1963, after a gap of 45 years, R. P. Kerr achieved the next significant step by determining the space-time geometry outside a *rotating* object of mass M and angular momentum S. Finally in 1965 E. T. Newman and his collaborators solved the problem of space-time geometry outside an object of mass M, electric charge Q, and angular momentum S. This solution describes the most general type of black hole which Price's arguments suggest as the final state of a collapsing body. We will now discuss the geometrical properties of this black hole – known commonly as the *Kerr–Newman black hole*.

It is convenient to describe these properties in terms of three length-scales associated with the mass, the angular momentum, and the electric charge:

$$R_s = \frac{2GM}{c^2}; \quad a = \frac{S}{Mc}; \quad q = \frac{\sqrt{G}Q}{c^2}.$$

For the Schwarzchild black hole $a = 0$, $q = 0$; for the Kerr black hole $q = 0$, while for the Reissner–Nordstrom black hole $a = 0$.

Like the Schwarzschild black hole, the Kerr–Newman black hole has an *event horizon* which is spherical in shape. Its surface area A is given by the formula

$$A = 4\pi(r_+^2 + a^2)$$

whereas its 'radius' is given by

$$r_+ = \tfrac{1}{2}\left[R_s + \sqrt{R_s^2 + 4a^2 - 4q^2} \right].$$

Notice that in Euclid's geometry the surface area of a sphere of radius r_+ is simply $4\pi r_+^2$. That A differs from this value is an indication of the fact that the geometry of the black hole is *non*-Euclidean.

There is another surface of physical significance surrounding the horizon. This surface is known as the *static limit* and it is not spherical. It is bun-shaped, being flattened at the poles which lie on the axis of rotation of the black hole. Figure 4.2 shows the equatorial and meridional sections of

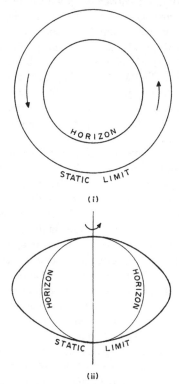

Fig. 4.2. The equatorial and meridional sections of a Kerr–Newman black hole are shown respectively in (i) and (ii). The ergosphere is the region whose inner boundary is the horizon and the outer boundary the static limit.

the black hole. These show that the surface of static limit touches the horizon at the poles.

What is the significance of the static limit? To understand this consider the example of the rotating Earth. The direction of the Earth's rotation is from west to east (around the north–south axis) and the effect of this rotation is felt not only by bodies on the Earth but also by bodies in the air. For example, birds flying up in the air and down again return to their starting places – they do not find the Earth's surface to have shifted eastwards while they were in mid-air. In fact, this observation was one of the major arguments by the Aristotelians against the rotation of the Earth. What in fact happens is that the atmosphere (in which the bird flies) is *carried along* with the Earth and therefore the bird does not find any relative displacement with respect to the Earth.

One of the consequences of general relativity is that the space-time around a rotating object is *carried along* with it, in a fashion somewhat analogous to the example given above. To test this effect on the space-time around the Earth an experiment has been proposed which involves launching a gyroscope in orbit round the Earth. Normally the axis of a gyroscope is fixed in space. However, if the above mentioned general relativistic effect is present, the axis should precess about a fixed direction. (Precession means moving in such a way that the axis describes a cone about a fixed direction.) Present technology has reached the stage where the predicted precession rate of $\sim 7''$ per year at a height of 800 km can be measured.

However, to return to our rotating black hole, which also carries its surrounding space-time along with it, the effect dies out with distance. So the effect is negligible with respect to a distant star. If we consider the scenario in Fig. 4.3 we have an observer O in the equatorial plane of the black hole and S is the distant star. The continuous arrow shows the tendency of the black hole to carry O along in the direction of rotation. To maintain its direction fixed relative to S, O has to apply a force (shown by the dotted line) in the opposite direction to counteract this tendency. The magnitude of the required force increases as O approaches the black hole: it becomes infinite when O arrives at the static limit. At and beyond the static limit the observer cannot maintain his position fixed relative to the distant star no matter how hard he tries. He is inexorably swept along by the black hole in the direction of rotation.

Notice that, unlike the horizon, the static limit does *not* prevent outward leakage of information. Material particles, photons, etc. *can* come out from the region between the horizon and the static limit – a region known as the *ergosphere*. We will return to the significance of the ergosphere later in the chapter.

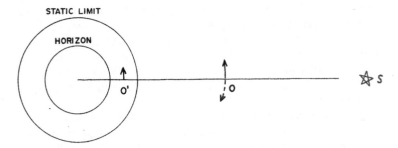

Fig. 4.3. The static limit is shown in its equatorial cross section. The observer O outside this limit can align himself in a fixed position relative to distant stars like S by exerting a suitable force to counter the tendency of the black hole to carry the observer in the direction of its rotation. The observer O' in the ergosphere is inexorably carried along by the rotating black hole.

The Laws of Black Hole Physics

During 1969–73 considerable progress was made in our understanding of how black holes behave in relation to ordinary matter. The physical behaviour of black holes can be summarized with the help of four laws. For a reason to be stated later, these laws have been numbered from zero to three. It is, however, more instructive to follow the development of these laws in the chronological sequence of their discovery than in the above numerical order.

The first law of black hole physics is simply a collection of the well known conservation laws in physics–the laws of conservation of matter and energy, conservation of momentum, of angular momentum, of electric charge, and so on. Since the black holes are products of ordinary matter and are subject to the same basic laws of physics which govern matter in general, it is hardly surprising that they should follow the first law. For example, if a black hole has mass M and it absorbs a quantity of matter of mass m and energy of quantity E, its mass will increase to

$$M' = M + m + \frac{E}{c^2}.$$

The last term on the right hand side is simply the mass equivalent of energy E as per the Einstein relation

Energy = Mass $\times c^2$.

This example raises an interesting question which brings us to the second law of black hole physics. The question is: 'can we proceed in the direction opposite to what the above example illustrates, i.e. can we decrease the mass of the black hole by extracting energy from it?

The Schwarzschild black hole clearly answers this question in the negative. Its horizon at $R = R_s$ is a one way membrane: it allows matter or radiation to fall in but does not allow it to escape. The black hole mass can therefore only increase – it can never decrease. No energy extraction is possible in the case of the Schwarzschild black hole.

The Kerr–Newman black hole, however, does allow energy extraction. An explicit process by which this can be achieved was first outlined by Roger Penrose in 1969. The Penrose process is described in detail in Box 4.1. What this process achieves is a simultaneous reduction of the mass and

Box 4.1 The Penrose process

The process proposed by Penrose is in fact a thought experiment designed to extract energy from the rotating Kerr black hole by using the properties of the ergosphere discussed in the text. As shown in Fig. 4.4, the process involves dropping a mass into the ergosphere, arranging for it to split into two bits, with one bit falling inside the horizon and the other escaping outside the ergosphere.

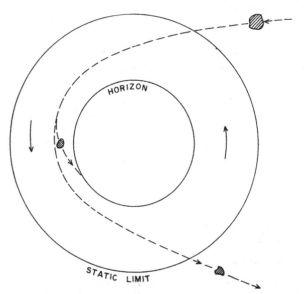

Fig. 4.4. In the Penrose process the energy of the incoming projectile is less than that of its part that emerges from the ergosphere.

What happens here is that the mass entering the ergosphere is made to rotate along with the black hole, as discussed in the text. It therefore acquires energy as well as angular momentum from the black hole. When it splits and part of it falls into the black hole, it loses a fraction of the acquired energy and angular momentum back to the black hole. The escaping portion *can*, however, emerge with so much energy that it exceeds the energy of the total mass that originally went into the black hole!

With the help of the second law of black hole physics we can understand the limitations on the Penrose process. Note that, as we reduce the mass and the angular momentum of the black hole, we can at best keep its area A constant. In Fig. 4.5 we see that if we travel along a constant area curve we finally end with the Schwarzschild black hole of zero angular momentum. If the *third* law of black hole physics holds, then we can at best start this process when the Kerr black hole is in what is known as the *extreme* state (when its surface gravity is zero). Here its angular momentum is so large that $R_s = 2a$. Let us denote by M_0 the starting mass of the black hole in this extreme state. Then its surface area is (by the formula given in the text)

$$A_0 = 2\pi R_s^2 = 8\pi \frac{G^2 M_0^2}{c^4}.$$

In the final state its mass is M_i, say. The area of the Schwarzschild black hole of this mass is

$$A_i = 16\pi \frac{G^2 M_i^2}{c^4}.$$

By the second law of black hole physics A_i cannot be less than A_0. *At best* we may equate A_0 to A_i giving

$$M_i = \frac{M_0}{\sqrt{2}}.$$

Thus the Penrose process can at most extract $(M_0 - M_0/\sqrt{2})c^2$ of the original mass energy $M_0 c^2$. The available energy is therefore $(\sqrt{2}-1)/\sqrt{2}$, i.e. nearly 30 per cent of the total energy. (By contrast, the hydrogen fusion process yields only ~ 0.7 per cent of the total energy.)

The mass M_i is known as *the irreducible mass* of the Kerr black hole.

the angular momentum of the black hole. It is also possible to reduce the mass of the black hole if we also simultaneously reduce its electric charge. In either case (or in a combination of both) the black hole yields energy for external use. The ergosphere of the rotating black hole plays an important role in the Penrose process. This is the region near the black hole where it is possible to get useful *work* out of the black hole (*ergon* in Greek means work).

The Penrose process illustrates how far it is possible to extract energy from a black hole. The process clearly cannot operate when all the angular momentum and electric charge have been extracted from the black hole, and it has become a Schwarzschild black hole. These limitations are part of a general rule which is known as the second law of black hole physics.

The *second law* states that in no physical process can the total surface area of all participating black holes ever decrease. Let us look at the Kerr–Newman black hole in the light of this law.

The area of this black hole, as we saw earlier, is given by

$$A = 4\pi(r_+^2 + a^2)$$

$$= 2\pi R_s \left\{ R_s + \sqrt{R_s^2 - 4q^2 - 4a^2} \right\} - 4\pi q^2.$$

In any physical interaction with the black hole we cannot decrease A. At best we can keep A a constant. In Fig. 4.5 we see what the $A = $ constant

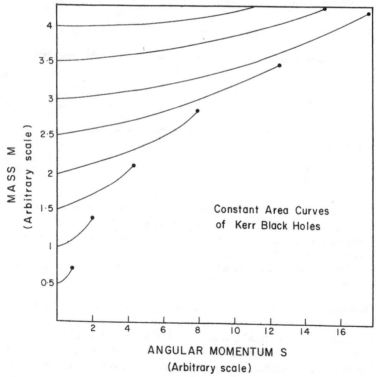

Fig. 4.5. Along each of the above curves the area of the Kerr black holes is the same. Higher curves in the figure correspond to greater areas.

curves look like when the mass M is plotted against the angular momentum S of a Kerr black hole. (To simplify the discussion we have taken $Q = 0$.) If we perform our energy extraction process most efficiently the area of the black hole stays constant, and we can stay on a particular curve of Fig. 4.5. To extract energy from the black hole we have to move from right to left. If the process is not 100 per cent efficient we have to jump to a higher curve of larger A-value. Clearly the extent of available energy from the black hole has decreased in this way.

These ideas have a striking similarity to *thermodynamics*, the science which describes the relationship of heat energy to mechanical work. The steam engine in the last century demonstrated for the first time how heat energy can be converted into other forms suitable for mechanical work. The first law of thermodynamics states that the amount of work so obtained is equal to the heat energy spent. In other words, the conversion from heat to work is done subject to the law of conservation of energy. The heat engines (of which the steam engine is one type), however, clearly showed that this conversion cannot be done to an arbitrary extent, nor very efficiently. There is considerable wastage (again in the form of heat loss) of the original heat energy, and only a small part is delivered as mechanical work.

The work of the nineteenth century physicists Clausius, Kelvin, Boltzmann, and others clarified the situation in terms of the now well known laws of thermodynamics, of which the first law has just been stated. The second law introduced the property called *entropy* which measures the disorderliness of a physical system. The second law states that an isolated system behaves in such a way that its entropy can never decrease. At best, the entropy remains constant. It was this second law which placed limits on how much mechanical energy could be extracted from a heat engine, and how efficiently.

The analogy between energy extraction from a heat engine subject to the second law of thermodynamics and the energy extraction from a black hole subject to the second law of black hole physics is now clear. The area of a black hole plays the role analogous to that of entropy of a physical system. Can this analogy be carried further?

The answer is 'yes', subject to certain limitations. Apart from entropy, another property plays an important role in thermodynamics. This is the property of *temperature*. Temperature measures the extent of small scale dynamical activity in a hot body. We say that a body 'feels hot' because its atomic and molecular constituents are in rapid random motion. The more rapid this motion, the hotter the body feels. The 'hotness' is measured by the temperature. Mathematically, temperature is proportional to the energy per constituent of the body, and the statement that 'the body as a whole has a temperature' means something only when all its constituents have settled down to an equilibrium state, i.e. when, by and large, there is no net give or take of energy between the different constituents of the body. This is known as the state of *thermal equilibrium*.

In black hole physics a corresponding state exists for axisymmetric stationary black holes. Stationarity implies a state which does not change with time and axisymmetry implies symmetry about some axis, e.g. the axis of rotation. In such a state the black hole possesses a property whose quantitative measure is the same all over the event horizon. This property is known as the *surface gravity* of the black hole, and it is usually denoted by κ. As its name implies, κ measures how fast the particles are accelerated towards the black hole (in analogy to g, the acceleration due to gravity on the surface of the Earth, which we encountered in Chapter 2). For a Schwarzschild black hole of mass M, the surface gravity is given by

$$\kappa = \frac{c^4}{4GM}.$$

The existence of a uniform temperature for a body in thermal equilibrium is the *zeroth law* of thermodynamics. The corresponding zeroth law of black hole physics establishes the existence of a surface gravity which is constant over the horizon of an axisymmetric stationary black hole.

In Box 4.2 we show the analogy between thermodynamics and black hole physics with the help of the Kerr–Newman black hole. The analogy shows through not only in the qualitative behaviour discussed in the text but also in the quantitative mathematical description of Box 4.2.

Box 4.2 The analogy between thermodynamics and black hole physics

Suppose we make small changes in the mass (M), angular momentum (S), and the electric charge (Q) of the Kerr–Newman black hole. As stated in the text this will result in a change of its surface area also. Simple calculus tells us that this change can be described by the following differential relation:

$$\delta Mc^2 = \kappa \frac{\delta Ac^2}{8\pi G} + \frac{a\delta Sc^2}{a^2 + r_+^2} + \frac{r_+ Q\delta Qc^2}{a^2 + r_+^2}.$$

We remind the reader that the area A and the surface gravity κ are given by

$$A = 4\pi(a^2 + r_+^2) = 2\pi(R_s^2 + R_s \sqrt{R_s^2 - 4q^2 - 4a^2}) - 4\pi q^2;$$

$$\kappa = \frac{\sqrt{R_s^2 - 4q^2 - 4a^2}}{2(a^2 + r_+^2)}.$$

We compare this relation with the thermodynamic relation

$$\delta U = \theta\delta\Sigma - p\delta V$$

where θ is the temperature, Σ the entropy, and U the internal energy of the system. p and V are the pressure and volume of the system (often in the form of a fluid). This relation relates the increase in the internal energy of the system to the change in entropy and to the work done by (or against) the pressure. Thus if pressures put in work and compress the system so that the change in the volume δV is negative, this work increases the internal energy of the system.

Now read these two relations in conjunction with the second laws of black hole physics and of thermodynamics:

$$\delta A \geqslant 0, \quad \delta \Sigma \geqslant 0$$

and the analogy becomes clear. The area A is analogous to entropy Σ, the surface gravity κ is analogous to the temperature θ, and the work done in changing the angular momentum or the electric charge of the black hole is analogous to the work done in changing the volume of the thermal system. In each case the net result of the two relations is to change the energy of the black hole or of the thermal system.

Note that the Hawking process makes the analogy *exact* in the sense that the *qualitative* similarity of κ to θ and of A to Σ can be made *quantitative*. It is easily verified that with the formulae given on page 72

$$\theta \delta \Sigma = \frac{\kappa \delta A}{8\pi G} c^2.$$

The third law of black hole physics makes it impossible to attain $\kappa = 0$ through a finite series of operations. $\kappa = 0$ for a Kerr–Newman black hole implies the *extreme* state when the horizon disappears and the singularity becomes naked.

Finally we come to the *third law* of black hole physics, which states that under no finite system of physical processes can the surface gravity of a black hole be made zero. This is analogous to the third law of thermodynamics which states that the absolute zero of temperature can never be attained by any finite system of physical processes.

At the absolute zero of temperature all the random motion of the constituents of the body becomes zero. Although advances in technology have enabled physicists to achieve temperatures as low as a few microkelvins (10^{-6} K) on the absolute scale,* the absolute zero *per se* has eluded them. This is not surprising in view of the third law. What does $\kappa = 0$ mean? A Kerr–Newman black hole with $\kappa = 0$ represents a limiting state beyond which the Kerr–Newman solution has no horizon! It therefore reveals what is inside. Towards the end of the chapter we will

* The absolute scale of temperature is the centigrade scale shifted through $273°$ C. That is, the zero on the absolute scale is $-273°$ C. It is the absolute scale of temperature that naturally arises from thermodynamic discussions.

return to the intriguing question of what lies inside the event horizon of a black hole.

Do Black Holes Radiate?

The analogy between black hole physics and thermodynamics is very striking and highly suggestive. Does it imply that we take the correspondence

surface area ~ entropy
surface gravity ~ temperature

more seriously than just as an analogy? If we actually interpret the surface area of a black hole as giving in some way a measure of its disorderliness, then we are also forced to accept the surface gravity as giving a measure of the hotness of the black hole. But can a black hole be hot?

Thermodynamics tells us that a hot body radiates if it is placed in cooler surroundings. Can black holes radiate? The very definition of a black hole requires the answer to this question to be in the negative. Yet in early 1974 Stephen Hawking found a way to answer this question in the affirmative. Hawking's arguments considered how a black hole would behave quantum mechanically rather than classically. To see how a black hole can radiate we first examine the peculiar way the so-called vacuum behaves according to the rules of quantum theory.

In classical physics the 'vacuum' implies absence of everything: the classical vaccum contains nothing. This is an idealized state of a region of space, which in practice can be approximated to by actual experiments. However, in quantum physics even the ideal vacuum has non-trivial properties. The quantum vacuum is a swarm of particles and anti-particles* which are constantly being created and destroyed. These particles and antiparticles are considered to be *virtual* rather than *real* in the sense that they do not last long. In Fig. 4.6 (a) we see how the 'empty' space-time of quantum physics is made up of such virtual pairs. The *net* effect of these pairs is zero, but their existence represents fluctuations about the 'zero state' which are important in that they do influence physical processes. The importance of vacuum fluctuations in elec-tromagnetic processes has long been experimentally confirmed.

The Hawking effect is based on how these vacuum fluctuations interact with a black hole, and it is illustrated in Fig. 4.6(b). The space-time diagram shows the boundary of the black hole by a pair of thick lines. The gravity of the black hole acts on the virtual pairs and it could produce the

---

* Particles are made of matter and antiparticles of antimatter. A pair of particle–antiparticle cannot last long, as the two annihilate each other.

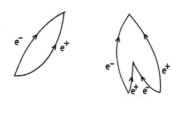

(a)

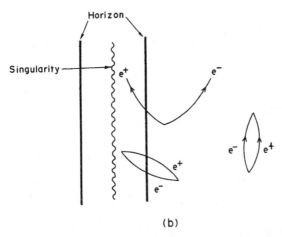

(b)

Fig. 4.6. The vacuum in flat space-time has virtual pairs being created and annihilated all the time as in (a) above. The effect of the black hole gravity is shown in (b).

different types of effects shown in Fig. 4.6(b). Both the members of the pair may fall into the black hole, or one member may fall and the other escape; or the pair is annihilated as it would be in the absence of the black hole. It is the second effect which is of interest. To a remote observer receiving the escaped member of the pair the impression will be created that the black hole has produced this particle (or anti-particle).

To decide which possibility actually takes place it is necessary to perform a quantum mechanical calculation. Such a calculation tells us the relative chances of occurrence of these various possibilities. The net effect is that the black hole appears to create particles at a certain rate. Hawking did such a calculation first for spin-zero particles. The calculation can, however, be done for particles of arbitrary spin. Although we have described the Hawking process qualitatively for particles of spin $\frac{1}{2}$, $\frac{3}{2}$, . . . (which have anti-particles), the process also works for particles of integral spin (1, 2, 3, . . .). In either case the outcome is remarkable.

The black hole radiates as if it were a black body!

A *black body* is an idealized enclosure containing sources and absorbers of radiation. The enclosure is idealized in the sense that no radiation is allowed to escape. Under such conditions the radiation eventually reaches an equilibrium state wherein its carrier particles are distributed in numbers according to their energies. For example, a heated gas oven with insulating walls has a near-black body distribution of electromagnetic radiation. The distribution is characterized by a single parameter, the temperature. Thus, in the case of electromagnetic radiation, its carrier particles, the photons, are distributed according to the following form:

$$N(E) = \frac{8\pi}{c^3 h^3} \frac{E^2}{e^{E/k\theta} - 1}.$$

Here $N(E)$ denotes the number of photons of energy E per unit energy band per unit volume in a black body distribution of temperature θ. The constant k, known as the *Boltzmann constant*, plays a key role in thermodynamics. The constant h is called Planck's constant. In fact the above black body distribution was first obtained by Max Planck in 1901 and is often called the Planckian distribution.

Hawking's calculations led him to the conclusion that the number of photons of energy E per unit energy band per unit volume emitted by a black hole is given by

$$N(E) = \frac{8\pi}{c^3 h^3} \frac{E^2}{e^{4\pi^2 E/\kappa h} - 1}.$$

Notice the similarity between the two distributions. It was this similarity which led Hawking to conclude that a black hole radiates as a black body and its surface gravity κ endows it with a temperature

$$\theta = \frac{h}{4\pi^2 k} \kappa.$$

The physical behaviour of surface gravity and temperature is not just analogous but it is *identical*. A similar relationship exists between the area of the black hole (A) and its entropy (Σ)

$$\Sigma = \frac{\pi k}{2hG} A.$$

In this way thermodynamics is shown to include the laws of black hole physics. The fact that a black hole radiates need not be considered a contradictory statement in view of the fact that we have changed the terms of reference from classical to quantum physics. There are several instances in which situations deemed impossible in classical physics are rendered probable in quantum physics. The so called '*tunnel effect*' is one such instance. Here the motion of an electron through regions of high potential barriers (i.e. regions of strong repulsion) is considered impossible according to classical Newtonian dynamics. The electron nevertheless tunnels through such regions with non-zero probability – a result which can only be understood within the framework of quantum mechanics.

Some of the details of the Hawking process are discussed in Box 4.3. It is worth mentioning here that the black holes discussed by astrophysicists

---

Box 4.3 The Hawking process

Taking the result that a spherical black hole of mass M radiates as a black body of temperature

$$\theta = \frac{\hbar c^3}{8\pi GkM} \cong 6 \times 10^{-8} \left(\frac{M_\odot}{M} \right) \text{K}$$

we now see how rapidly a low mass black hole evaporates.

The rate of mass loss is given by the differential equation

$$\frac{\mathrm{d}Mc^2}{\mathrm{d}t} = -\frac{1}{4} ac\theta^4 \frac{16\pi G^2 M^2}{c^4}.$$

This equation simply states that the energy lost by the black hole is that by a black body of surface area $16\pi G^2 M^2/c^4$, a being the radiation constant. Substituting for a in terms of $\hbar$ and c, we finally get the result that M becomes zero in a time

$$t = \frac{5120\pi G^2}{\hbar c^4} M^3 \cong 10^{76} \left(\frac{M}{M_\odot} \right)^3 \text{seconds.}$$

Hawking argued that black holes which formed soon after the big bang (see Chapter 8) would have evaporated by now if their time of evaporation was less than the age of the Universe, estimated at $\sim 4 \times 10^{17}$ s. The above formula tells us that primordial black holes of mass $\sim 10^{11}$ kg would now be in the final stages of evaporation. These evaporations would represent extremely hot black bodies which radiate in gamma rays. It does not seem, however, that the detailed features of the gamma ray bursts mentioned in Chapter 1 can be accounted for by the Hawking process.

---

are so massive that they hardly radiate even through the Hawking process. A Schwarzschild black hole of mass $6 M_\odot$ has an effective radiating temperature of only 10^{-8} K and the temperature of a more massive black hole falls further in inverse proportion to its mass. Thus a black hole of mass $60\,M_\odot$ will have one tenth of the temperature of the black hole of mass $6\,M_\odot$, i.e. a temperature of a nanokelvin.

However, if black holes of much lower mass existed, they would be much hotter. As shown in Box 4.3, the lower the mass of the black hole, the faster it radiates and so the shorter its life time. For a radiating black hole may eventually lose all its mass! In the final stages of such a process the radiation is so rapid that astrophysicists use terms like 'evaporating black holes' or 'exploding black holes' to characterize these events. We will return to the astrophysical implications of these mini black holes in Chapter 8 (see also Box 4.3).

The End-point of Gravitational Collapse

We began this chapter with a discussion of gravitational collapse and we end it with the same topic. So far as the external observer is concerned, the future of the collapsing body is hidden from him as the body enters the event horizon. This does not mean, however, that the body itself has no future. It does have a future but not a pleasant one!

How will a man stationed on the collapsing body react to his surroundings? As the body shrinks he will notice the effects of gravity in a way different from that of the remote observer. If the remote observer sends him flashes of radiation, this radiation suffers spectral shifts of two opposing kinds. First there is the Doppler *red-shift* since the receiver (on the collapsing body) is receding from the source of light. This is, however, countered by the gravitational *blue-shift*. The net outcome, i.e. which effect wins out, depends on the details of the collapse. The man on the collapsing body does, however, continue to receive radiation from the remote observer, even after the body has crossed the horizon.

The unpleasant effects of gravity, however, begin to show up in a different way, through its *tidal* interaction. To understand this effect consider what happens when a man stands on the surface of the Earth. Newton's inverse square law of gravity tells us that the man's feet are attracted towards the Earth with a greater acceleration than his head is. The difference in accelerations is so small as to be negligible. (The difference is of the order of four millionth part of g.)

For the collapsing body, however, this difference in the accelerations is very large. And as the man stands on the collapsing body he will find that his feet are rushing ahead far faster than his head, with the result that his whole body is stretched. A similar stretching force is exerted on the Earth by the Moon's gravity and this force causes tides in the oceans. Hence the gravitational stretching force is usually called the *tidal force*.

As the body collapses the tidal force grows rapidly. For a body of solar mass $M_\odot$, the tidal force is so large that the man is torn apart long before the surface of the body reaches the event horizon! This is just as well, for if the man had survived with a super-human effort, he would have encountered an even more bizarre state of affairs.

For, in a finite time kept by clocks stationed on the collapsing body, the entire body shrinks to a point. At this point the space-time is infinitely curved, the tidal forces are infinitely large, and the matter density is infinite. Such a state is known as a '*space-time singularity*'. Physical laws and their mathematical description break down at this singular epoch and to all intents and purposes this event is regarded as marking the end to the future of the collapsing body (and of the unfortunate man stationed on it).

Is such an abrupt end to the future inevitable? Must singularities exist at the end of gravitational collapse? The answer is 'yes' – provided classical physics and Einstein's general relativity are supposed to be valid right to the end. This is the sum and substance of the work of Roger Penrose, Stephen Hawking, and Robert Geroch who examined various situations of strong gravity in the mid 1960s.

Recent work by the author shows, however, that this outcome may be avoided in quantum physics. The quantum fluctuations of space-time geometry in a collapsing object grow as gravitational effects grow. And, like the tunnel effect, it is not improbable that the object may bypass the space-time singularity altogether. Even within the framework of classical relativity the space-time singularity may be averted if the physical behaviour of matter turns out to be drastically different from what we know so far from laboratory experiments.

Another related question is whether the space-time singularity is visible to the external observer. No definite answer is known to this question, although indications are that an event horizon always develops before the singular state is reached. The assumption that the external observer is prevented from seeing the singularity by the intervening horizon has been termed by Penrose as the hypothesis of *cosmic censorship*. The validity of the second and third laws of black hole physics is based on the validity of the cosmic censorship hypothesis. If a singularity unshrouded by a horizon does exist, it will be called a *naked singularity*.

This ends our discussion of black holes as creatures of general relativity. They have certainly enriched the theoretical understanding of general relativity and of the way physical processes should operate in highly non-Euclidean space-times. The proof of the pudding, however, lies in the eating. The reader may well ask: 'is the black hole pudding of any use to the astrophysicist?' In the remaining part of this book we will encounter several instances in which these theoretical speculations have helped in our understanding of the violent phenomena in the universe.

5 Star Explosions and their Aftermaths

If there be the effulgence of a thousand suns bursting forth all at once in the heavens, even that would hardly approach the splendour of the mighty Lord.

The Bhagavadgita XI-12

Exploding stars called supernovae are rare events. In our Galaxy there are around a hundred thousand million stars. The rate at which new stars are being born is estimated to be about 1–10 per year. By contrast supernovae occur probably not more than twice or three times per century. If we take into account the enormous size of our Galaxy (a flattened disc shaped object of diameter $\sim 100\,000$ light years and a thickness of ~ 6000 light years) we are not likely to see most of the supernovae in it. For the interstellar dust tends to obscure the light from them. In fact, as mentioned in Chapter 1, since the Crab Nebula supernova was seen to explode in 1054 A.D., only two other supernovae have been seen in our Galaxy. Nevertheless, supernovae can be detected, even after they have become extinct, through their effects on the surrounding interstellar medium and through the remnants that are left after the explosions.

Why do stars explode? Does every star explode? What is the form of the debris in a stellar explosion? Is anything coherent left behind after the explosion?

These are the questions we will try to answer in this chapter. Their answers lead us not only to a better understanding of the phenomenon of star explosion but they also offer clues to other interesting phenomena such as the origin of the Solar System, star formation in general, pulsars, cosmic rays, etc.

How do Stars Evolve?

The questions raised above cannot be answered without an adequate understanding of stellar structure and evolution. An explosion indicates a breakdown of internal equilibrium, and to understand why and when such a breakdown of equilibrium is likely to occur it is necessary to know in the first place how this equilibrium is maintained in stars in general.

In the previous chapters we have seen how the self-gravity of a massive object tends to make it shrink. Unless there were internal pressures of the required magnitude to withstand this shrinking tendency, the massive object would collapse. The fact that the Sun, with a mass as large as 2×10^{30} kg, can maintain a steady size is proof of the existence of strong internal pressures within it. How did the Sun acquire these pressures?

The current ideas on star formation tell us that stars are formed by the shrinkage of an interstellar cloud of gas and dust. As a cloud shrinks it tends to fragment into a large number of smaller units. Each unit contracts further and in this process begins to get hotter in the central regions. These early stages of a star's life were discussed by the Japanese astronomer Chushiro Hayashi. When the central temperature reaches a high enough value the process of nuclear fusion is triggered off and the star is said to have 'come of age'.

For example, the Great Nebula in Orion, shown in Fig. 5.1, is believed to be a site for star formation in the recent epochs. This nebula is located at a distance of about 1500 light years and it contains a large number of *proto* stars. In proto stars the internal temperatures have not reached high enough values to trigger off thermonuclear reactions. These proto stars, because of their temperature, however, do radiate energy largely in the infrared part of the electromagnetic spectrum. The detection of infrared emitters in the Orion Nebula therefore supports the view that stars are being born there.

Nevertheless there is a problem associated with the initial stages of star formation which we will discuss later in this chapter. It is a problem to which the supernovae provide an answer.

Once a star begins to function as a nuclear reactor, the qualitative features of its evolution are briefly as follows. In the beginning, the nuclear reactions convert hydrogen to helium by the process of *fusion* of nuclei (see Appendix D). This process releases energy which keeps the star in equilibrium against its self-gravity. While this reaction is going on the star is said to be on the *main sequence*. The main sequence stage is the *longest* stage in the star's life, its overall period depending on the total mass of the star. The larger the mass, the shorter is this period, since the rate at which hydrogen is fused is faster for more massive stars.

When hydrogen is exhausted, especially in the inner core of the star, the core begins to contract, since with the stoppage of nuclear reactions there is no force strong enough to combat gravity. However, as the core contracts, it heats up further and this rise in the core temperature triggers off the next series of nuclear reactions. In these reactions, helium is converted to carbon, then carbon to oxygen, then to neon, and so on. At each stage of this series, a nucleus is made bigger by the addition of an alpha particle (i.e. a nucleus of helium). This process increases the charge of the nucleus by 2 and its mass by 4. Whenever one nuclear species is converted to the next higher one in this way, the fusion process is stopped. This stoppage leads to the weakening of the resistance to gravity which results in a contraction of the core and the rise of central temperature. When the core temperature rises to a high enough value the next nuclear reaction takes place which, while it goes on, prevents a further contraction of the core. This reaction builds up the next higher nucleus in the alpha particle ladder. At a high enough temperature even bigger nuclei can be fused together. For example, in carbon burning two carbon nuclei are fused to give a magnesium nucleus. And so this stop-go process goes on . . .

Yet there is a limit to how high we can climb this ladder in this fashion. As discussed in Appendix D, there is a limit to the size of a nucleus which can be held in a stable form by the nuclear forces. This limit is reached in the so-called iron group of nuclei–which consist of iron, cobalt, and nickel–with atomic mass 56.

What has happened to the star while these reactions are going on? The specific details of the answer to this question depend, again, on the mass of the star. In general, the star's core progressively contracts and gets hotter while the outer envelope expands and cools. Thus to an outside observer the star appears to become larger in size and redder in appearance (the visible effect of cooling). Such stars are known as *red giants*.

However, this is the stage when the star is ripe to become a supernova, provided it is massive enough. Since our interest here is in supernovae, we will henceforth concentrate on the evolution of more massive stars.

Supernova Explosion

Present-day calculations of stellar evolution do not pinpoint an exact mass limit beyond which a star becomes a supernova. Roughly, this limit is around $6\,M_\odot$. That is, a star six or more times more massive than the Sun becomes a supernova and explodes. To see how explosive conditions develop within a massive star, let us concentrate on the fate of a star of mass $25\,M_\odot$. The qualitative aspects of stellar evolution described above are summarized below in Table 5.1 for a star of this mass.

In this table, a typical row describes the core temperature and density when a particular element is undergoing fusion. Thus the first row tells us that for fusion of hydrogen the central density is 5 times that of water and the central temperature is 6×10^7 K. This reaction takes seven million years. (By contrast the Sun is expected to be on the main sequence for the much longer period of ten thousand million years.)

This information is shown in Fig. 5.2. One consequence of this evolution is the dramatic rise in the rate at which neutrinos come out of the

Table 5.1

Fusion stage	Temperature (K)	Density (in g/cm^3)	Time-scale
Hydrogen	6×10^7	5	7×10^6 years
Helium	2.4×10^8	700	5×10^5 years
Carbon	9.3×10^8	2×10^5	600 years
Neon	1.75×10^9	4×10^6	1 year
Oxygen	2.3×10^9	10^7	6 months
Silicon	4.1×10^9	3×10^7	1 day

Based on the calculations of S.E. Woosley and T.A. Weaver.

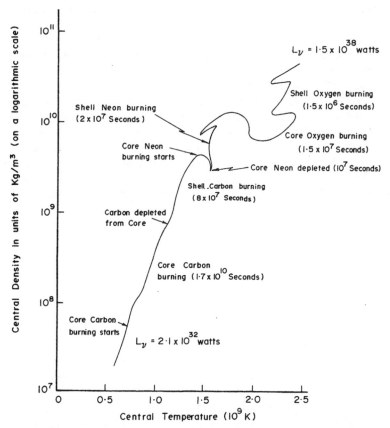

Fig. 5.2. Figure showing the evolution of the core of a 25 $M_\odot$ star, with the central density plotted against the central temperature. Some landmarks in the fusion process and the characteristic time-scales are also shown. Notice the rise in neutrino luminosity L_ν from the early to late stages, by a factor of nearly a million. (Adapted from the work of S. E. Woosley and T. A. Weaver.)

star, i.e. in the star's neutrino luminosity. Although initially the energy released by the nuclear reactions is mostly radiated by the star in the form of visible light, in later stages this energy loss is predominantly in the form of neutrinos. If a future technology can provide us with a sensitive neutrino detector, the dramatic rise in the neutrino emission from a star will give us the advance indication that the star is entering the supernova stage: that it is about to explode!

What is the composition of the star as a whole in this pre-supernova stage? This is shown in Fig. 5.3. The star has a shell-like structure, much like that of an onion. Each shell as we go inwards from the surface is progressively hotter and more advanced in the level to which fusion has taken place. Thus the outer shell is of hydrogen which has remained unchanged because the temperature in the outermost parts of the star never reached the fusion threshold. Next we have a shell consisting predominantly of helium, with a slight admixture of carbon. The next shell, known as the carbon shell, contains mainly carbon and also smaller amounts of oxygen, neon, and magnesium. There are thinner shells of neon, oxygen, and silicon to follow. The innermost part of the star contains, as mentioned earlier, the iron group of nuclei. Since these nuclei have been built up by addition of the alpha particle whose mass is 4 units (see Appendix D), the masses of these nuclei increase in steps of 4.

Now something happens in the central region of the star which triggers off the explosion. The exact details are not clear, although a broad outline is now beginning to emerge after a number of investigations. The process

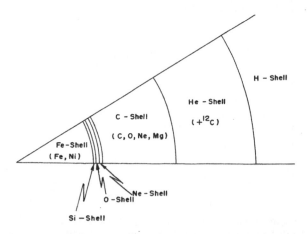

Fig. 5.3. The onion shell structure of the star in the pre-supernova stage. The shell sizes are not drawn to scale. The extent of the outer hydrogen envelope is not shown.

begins with the core of the star giving way to its gravity, its pressure being no longer enough to hold it in equilibrium. This happens because the iron group of nuclei marks the end to the process of energy generation by nuclear fusion – the process which had hitherto provided energy to withstand gravity.

As the core contracts it releases energy (see Chapter 2 for release of energy by gravitational contraction). Somehow this energy is extracted from the core and dumped in the outer regions of the star. The most dominant source of energy loss in the core is the breakup of the iron nuclei into alpha particles. What happens to the core henceforth is itself an interesting story to which we will return in the later parts of this chapter. At present let us concentrate on the outer parts of the star which have suddenly received a huge flux of energy from inside.

This process of energy dumping from the core to the envelope is often likened to the working of a pump. The piston of a pump is pulled back and then pushed forward to pump air into a balloon, say. In the same way the collapse of the core is followed by a bounce which gives a push on the envelope. (Why the core bounces, we will consider later when we discuss the future of the core.)

What we have described above is a likely scenario. The full quantitative details have still to be worked out. The outcome, however, is expected to be a push on the envelope from within. The push is in the form of a *shock wave* of gigantic dimensions. A shock wave typically represents a surface of discontinuity as shown in Fig. 5.4: that is, the pressure, density, and

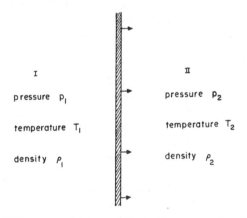

I

pressure p_1

temperature T_1

density ρ_1

II

pressure p_2

temperature T_2

density ρ_2

Fig. 5.4. In a typical shock wave, the wave front (shown by thin shaded section) separates the two regions I and II with differences in pressure, temperature, and density.

temperature of the gas on the two sides of the surface are not the same. This disturbance travels through the medium much like the sonic boom which is generated when a supersonic aircraft like the Concorde flies overhead. The giant shock wave travels through the star and produces two effects.

First it produces heating of the material in the regions outside the core to trigger fusion reactions in them, *albeit* for a short while. This process is termed *explosive nucleosynthesis*. The peak shock temperature reaches the high values of $\sim 4 \times 10^9$ K for a few tenths of a second in the silicon and oxygen cores, and lower values as the shock progresses outwards. Although in terms of production of nuclei the explosive nucleosynthesis is not as prolific as the nucleosynthesis which went on in the early stages of the star's life, its effects have been noticed in a dramatic fashion in our Solar System. We will discuss this topic after completing our description of the supernova explosion.

The second effect of the shock wave is to accelerate the matter in the envelope to velocities *beyond* the escape velocities (see Chapter 2 for a discussion of escape velocities). This process results in a disruption of the envelope, which is blown off into interstellar space. This is how the star finally explodes.

The visible effect of the explosion, as in the case of the Crab supernova, is to enhance the luminosity of the star briefly and then to slowly decrease it. At its peak brightness the supernova can outshine the entire galaxy, which may contain as many as a hundred thousand million ordinary stars! The typical supernova light curves are shown in Fig. 5.5. The detailed examination of the light curve of a supernova, coupled with the theoretical understanding of the process of explosive nucleosynthesis, can tell us a lot about what goes on inside the supernova prior to and after the explosion.

Supernovae and Star Formation

While the supernova marks the 'death' of a star in a certain sense, it leaves a lasting impression on the next generation of stars to follow. This is the exciting conclusion to come out of some detailed investigations of the long-standing problem of the origin of the Solar System and of star formation in general.

More than thirty years ago two leading astronomers, J. Opik and F. Hoyle, independently suggested that a gigantic star explosion may trigger off the process of formation of new stars. The reason for such a hypothesis was the difficulty faced by the theory of star formation which we have briefly referred to in the beginning of this chapter. Now is the appropriate moment to consider that difficulty and to see what solution a supernova is likely to offer.

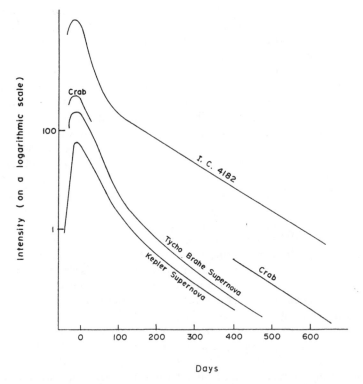

Fig. 5.5. The light curves of the three historic supernovae in our Galaxy and of the supernova seen in 1938 by W. Baade in the galaxy I. C. 4182 are shown. The intensity of light is plotted on a logarithmic scale against time (in days) on a linear scale. The early rise in intensity is associated with the energy released by the explosive nucleosynthesis. As the debris is hurled out, the light first rises and then falls rapidly, within 20–30 days. Thereafter the slow fall in intensity is connected with the fission and β-decay processes of heavy nuclei (see Appendix D). The curves are on a relative scale and do not reflect the actual intensities of the supernovae. A drop in intensity by a factor 100 is shown for comparison.

As mentioned earlier, stars are supposed to form out of a contracting cloud of gas and dust. The difficulty with this contraction hypothesis is that the interstellar clouds are very diffuse to begin with. In a diffuse state of matter the force of gravity is not strong enough to induce contraction. By Newton's inverse square law, it is clear that the self-gravity of a rarefied cloud will be rather small, small enough in any case for the cloud's internal pressures to provide an adequate counter to any gravity induced attempt at contraction. How then could the contraction proceed in the first place?

If an external agency providing initial compression of the cloud became available, then, as the cloud contracts, its gravity picks up strength and can subsequently take over as the contracting force. All that is needed is an initial triggering mechanism.

Such a mechanism is provided by the shock wave released in a supernova. In a typical supernova the shock wave ejects the outer envelope. The ejected matter initially travels outwards with a velocity of 10 000 km per second. As it collides with the ambient medium it slows down. After 1000 years or so, it will have travelled a distance of nearly four light years and would be moving at the speed of 1000 km per second. The oldest left over bits of the supernova may be found as far as 200 light years from the site of explosion and travelling with the slow speed of only 50 km per second. Such pieces of debris, known as supernova remnants, are seen in many places in our Galaxy.

What is of interest to the theory of star formation in the above process is the possibility that the shock wave which pushes the remnant from the parent supernova may provide the necessary compression to an interstellar cloud to set it on its way to star formation. This idea has recently received support in the observations of the supernova remnant associated with the region known as Canis Major R1. W. Herbst and G. E. Assousa calculated the age of the remnant from its expansion velocity and diameter to be about 800 000 years. The stars in the neighbourhood of this shell seem to be in the pre-main-sequence stage (i.e. in the stage when they have yet to switch on their central nuclear reactors). Herbst, R. Racine, and J. Warner have estimated the ages of these stars to be as low as 300 000 years. At this age these objects are among the youngest stars known to date! Thus there is circumstantial evidence to link the formation of these stars with the expanding shell in their vicinity. The estimate of how much energy is needed to set the shell in motion (around 10^{44} joules) also indicates that only a supernova could have released so much energy as it exploded.

It is of course possible for a massive star to release this much energy *without* becoming a supernova, provided other conditions are right. For example, if the surrounding interstellar medium is homogeneous, a massive O-type star* can generate in it a shell of the above type. However, near Canis Major R1 the medium is *not* homogeneous, nor is there any massive star at the centre of the shell. The supernova hypothesis does, however, receive support from the fact that a star is seen to be rushing away from the region with a speed significantly higher than that of other stars in the neighbourhood. This may well be the star that fired the shell in a supernova explosion. A directed explosion would generate a recoil in the

---

* Stars have been classified, according to their different spectra, into types O, B, A, F, G, K, M, R, N. Of these, O-type stars are massive (mass 10–30 $M_\odot$) and very hot (surface temperature $\sim$ 30 000–45 000 K) and their spectra contain predominantly ionized helium lines.

star, much like the recoil of the cannon after it has fired a ball. The observed velocity of the star does fit with the recoil hypothesis.

Let us leave the case of Canis Major R1 and move closer to home where another type of evidence links supernovae with star formation. The evidence to be discussed now strongly suggests that our own Solar System may owe its origin to a nearby supernova.

In 1969 a meteorite fell in the Mexican village of Pueblito de Allende. Now known as the Allende meteorite, this piece of matter in our Solar System has yielded the significant clues which link it to a supernova. These clues are referred to as isotopic anomalies. Isotopes of the same element are nuclei with the same number of protons but with different numbers of neutrons (see Appendix D for details). Isotopic anomalies imply differences in the proportion of the various isotopes in the meteorite compared to the overall average seen in the Solar System.

When a supernova explodes it throws the material (hydrogen, helium, carbon, oxygen, . . .) in the various shells of its envelope out into the interstellar medium. This contamination of the surrounding medium persists for a short while; eventually, however, these impurities get blended with the overall interstellar matter. Hence if significant time has elapsed since a supernova last exploded in a given region, the stars formed there would show a uniform isotopic composition. If, however, star formation took place relatively soon after a supernova exploded in the region, the stars and their accessories (planets, comets, meteorites, etc.) would show evidence of supernova contamination through an inhomogeneity of composition. Of course, the interpretation of the evidence is not as straightforward as it may appear at first sight. For the various isotopes of the nuclei which are thrown out by the supernova are not all stable. Some of them decay into other nuclei with characteristic life-times. Some isotopes are stable, some are not. It is the unstable isotope that complicates the interpretation of the evidence on chemical composition. Yet at the same time such isotopes provide interesting information about time-scales, as will be seen from the example of the Allende meteorite.

One of the isotopic anomalies in this meteorite relates to the enhancement of magnesium – of the isotope ^{26}Mg. This isotope has the atomic mass of 26. The observation that ^{26}Mg occurs in greater proportion in the Allende meteorite (compared to the overall Solar System average) indicates that some process has been at work to provide the extra ^{26}Mg in the meteorite. Now this could happen in many ways, but one particular process seems most likely – through the decay of ^{26}Al. ^{26}Al is a radioactive isotope of aluminium and it changes into ^{26}Mg by the process:

$$^{26}Al \rightarrow \, ^{26}Mg + e^+ + \nu.$$

That is, of the 13 protons and 13 neutrons in ^{26}Al, one proton is changed to a neutron with the emission of a positron and a neutrino. The resulting ^{26}Mg nucleus has 12 protons and 14 neutrons. This process has a half-life of 720 000 years. That is, if we started with 100 nuclei of ^{26}Al, half of them (i.e. 50) would have decayed to ^{26}Mg in 720 000 years. By contrast the aluminium commonly found on the Earth is made of the stable isotope ^{27}Al with 13 protons and 14 neutrons.

Why should the excess ^{26}Mg in the Allende meteorite have originated from ^{26}Al? The reason for believing in this theory lies in the observed correlation between the aluminium content (^{27}Al) and the excess ^{26}Mg found in each mineral grain of the meteorite. This correlation strongly suggests a link beween magnesium and aluminium, a link which is naturally explained by the above decay process. (It should be remembered here that the original ^{26}Al, according to this interpretation, would be quite rare; it would occur in the ratio of $1:20\,000$ compared to the stable ^{27}Al.)

How could the original ^{26}Al appear in the Allende meteorite? After considering various alternatives the astrophysicists have found its supernova origin the best explanation. While discussing the explosion process we referred to *explosive nucleosynthesis*. This happens when the shock wave travels outwards from the core of the supernova, momentarily heating the outer shells to high enough temperatures to effect transmutation of nuclei. It is believed that ^{26}Al is made in the carbon shell of the supernova. Notice that ^{26}Al does not fall in the sequence of nuclei built by adding the alpha particle. It is formed from ^{24}Mg (which does occur in this sequence) by additions of neutrons(n) and protons(p) which are present in abundance during the explosive nucleosynthesis. A typical reaction sequence leading from ^{24}Mg to ^{26}Al is:

$$^{24}\text{Mg} + \text{n} \rightarrow {}^{25}\text{Mg}$$
$$^{25}\text{Mg} + \text{n} \rightarrow {}^{26}\text{Mg}$$
$$^{26}\text{Mg} + \text{p} \rightarrow {}^{26}\text{Al} + \text{n}.$$

Other sequences also exist and have been studied in detail.

Now we recall the fact that ^{26}Al has a half-life of 720 000 years. Thus after these isotopes were created in a supernova and deposited in the interstellar medium their species would have died out in a matter of a few million years, leading to ^{26}Mg. These magnesium isotopes would have mixed freely with the interstellar medium and the 'signature' of ^{26}Al would have been lost, had the medium remained undisturbed for long after the supernova explosion. In other words, had the Solar System formed several million years after the supernova explosion, the initial inhomogeneities of supernova contamination would have been wiped out.

Fig. 1.1. The Crab Nebula. (Photograph from Palomar Observatory, California Institute of Technology.)

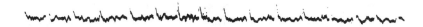

Fig. 1.2. A typical pulse pattern from a pulsar. The pulsar is PSR 1237 + 25. (By courtesy of the Radio Astronomy Centre, Ootacamund, India.)

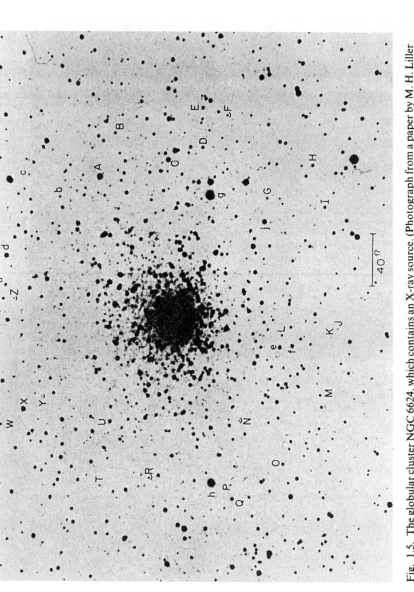

Fig. 1.5. The globular cluster NGC 6624, which contains an X-ray source. (Photograph from a paper by M. H. Liller and B. W. Carney in *Astrophysical Journal*, **224**, 385.)

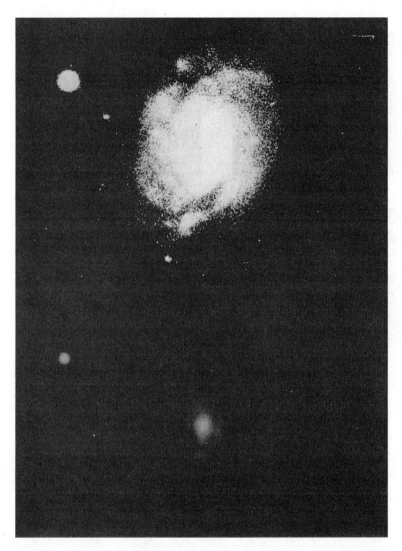

Fig. 1.7. The Seyfert galaxy NGC 1068. (Photograph from the Indian Institute of Astrophysics.) The lower half of the picture is underexposed to show the bright central nucleus.

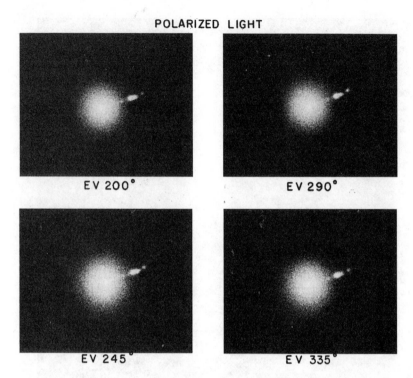

EV 200°

EV 290°

EV 245°

EV 335°

Fig. 1.8. Four views of the galaxy M 87 showing the jet-like structure from the centre. (Photograph from Palomar Observatory, California Institute of Technology.)

Although we have concentrated our discussion on the aluminium anomaly, there are also other isotopic anomalies in the Allende material. For example, the over-abundance of the oxygen isotope ^{16}O also gives a clear indication of its supernova origin. The fact that such inhomogeneities *are* seen in the Allende meteorite therefore strongly suggests that our Solar System began to be formed not long after a supernova exploded in its vicinity.

Of course this fact by itself does not *prove* that this supernova was responsible for the formation of our Solar System – just as the presence of the shell in Canis Major R1 does not *prove* that the young stars in its vicinity owe their origin to its presence. But when we take into account the rarity of supernovae ($\sim$ 2–3 per century) the above scenarios do begin to assume a highly suggestive form. Given the positive role supernovae are likely to play in triggering star formation the above examples cannot be dismissed as just coincidences.

Neutron Stars

We now leave the outer envelope of the supernova and return to the core in the middle. Recall that we left the core at the stage where it was collapsing under its self-gravity because its composition (in the form of iron group nuclei) no longer permitted further nuclear fusion which had hitherto provided pressure support against gravity. As the core contracts it heats up and the heavy nuclei in it begin to break up into small constituents like the alpha particles, neutrons, and protons.

Now in the laboratory experiments on the Earth a neutron does not last very long. In a matter of about 12 minutes* it decays into a proton, an electron, and an antineutrino:

$$n \to p + e^- + \bar{\nu}.$$

However, in the core of the supernova the high densities permit the neutron to exist as a stable particle. What is more, even the proton can turn into a neutron through reactions like

$$p + e^- \to n + \nu, \; p + \bar{\nu} \to n + e^+, \text{ etc.}$$

Thus the matter in the core begins to be more and more neutron-dominated and a swarm of neutrons begins to resist further compression of the core by means of a special kind of pressure known as the *degenerate neutron pressure*. This pressure builds up due to a quantum mechanical effect discussed in Box 5.1.

---

* This is the half-life of the neutron.

Box 5.1 The degenerate pressure of dense matter

One of the important results of quantum theory is that there exists a class of elementary particles called 'fermions' with the property that no two identical fermions can be found in the same state. Electrons, protons, and neutrons are fermions.

How do we say whether two particles are in the same state? The state of a particle is specified by a number of dynamical variables like energy, momentum, angular momentum, spin, etc. So if we consider the electrons in a given volume with a given energy, we would expect them to differ in their other dynamical properties, in this case, spin and momentum. The number of such electrons is limited by the above quantum theoretical result (known as Pauli's exclusion principle). The permissible number of electrons and their pressure increases with energy according to a specific law.

In white dwarfs or neutron stars the high matter densities make the above quantum result important. The electrons in white dwarfs and neutrons in neutron stars are distributed in such a way that all the low energy states are fully occupied and the pressure–energy relation follows the above mentioned law. The matter in all such cases is said to be *degenerate* because its constituents (the electrons in white dwarfs and the neutrons in the neutron stars) are severely restricted in their movements and other physical behaviour.

The building of this pressure in the core is a very sudden phenomenon: it takes less time than it has taken to write these few lines. The pressure in the collapsing star builds up to such an extent that it temporarily checks gravitational infall of matter and causes a bounce. This was the core-bounce that generated the shock wave in the supernova, referred to earlier.

The core, however, does not expand for long under the bounce since the pressure in it quickly falls. The core contracts again. It may in fact oscillate a few times and then settle down to a static shape, *provided* the degenerate pressure can successfully withstand the force of gravity. As mentioned in Chapter 4, this happens if the core mass is not excessive – in any case not more than 2–3 $M_\odot$ (the precise upper limit is still to be determined).

So it becomes necessary to know how much matter is left in the remnant core of a supernova. Calculations suggest, for example, in the case of a star of 25 $M_\odot$, the remnant may have a mass of the order of 1.6 $M_\odot$. Such a case therefore results in a static star made up predominantly of neutrons. The density of matter in the centre of a neutron star may be as high as 10^{15} times the density of water. A schematic diagram giving the composition of a neutron star of mass 1.4 $M_\odot$ and radius 16 km is given in Fig. 5.6.

The high densities inside a neutron star take us to the limit of what can be speculated about dense matter within the present-day knowledge of physics. It is therefore an intriguing question to ask what would happen if

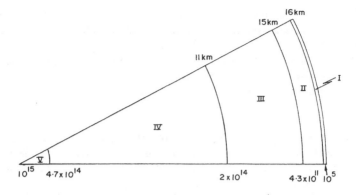

(The figures below the horizontal lines denote the densities in terms of the density of water)

Fig. 5.6. A schematic cross-section of a neutron star of mass 1.4 $M_\odot$ and radius 16 km. The outermost region (I) is a thin magnetic layer of small tightly bound atoms of density $\sim 10^4$–10^5 times the density of water and temperature $\sim 10^5$–10^6 K. The next region (II) has nuclei arranged in a body centred cubic lattice forming mostly a solid crust with density rising to $\sim 4.3 \times 10^{11}$ times the density of water. The inner crust (III) also has similar lattice structure with density going up to $\sim 2 \times 10^{14}$ times that of water. The extended region (IV) next is made up mainly of neutrons with the inner density going up to $\sim 4.7 \times 10^{14}$ times the density of water. In some models the density may go to twice this value. The innermost region (V) may consist of hyperons and pions whose behaviour is still largely unknown. The density here may rise to $\sim 10^{15}$ times the density of water.

a very massive star left behind a remnant core of mass $> 3\ M_\odot$ in a supernova explosion. From the above considerations such a remnant cannot exist as a static neutron star. The core must continue to collapse and reach even higher densities than those in a neutron star. Can matter at such densities provide an effective counter to gravity? The answer to this question takes us beyond the present frontiers of the physics of dense matter. If one must speculate on the conservative side, then the answer to the above question is negative. The core in that case undergoes a gravitational collapse to become a black hole as described in Chapter 4.

However, to get back to neutron stars! The idea of neutron stars had been suggested by Landau and by Oppenheimer and Snyder back in the 1930s. Their relation to supernovae and their place in the overall evolutionary sequence of a star became clear only in the 1960s. Since the neutron stars with an initial temperature of 10^{10} K cool down rapidly (in a matter of 100 years the temperature falls to 10^8 K and in a million years to $\sim 10^6$ K) it becomes a problem for the astronomer to detect them since, at the surface temperature of $\sim 10^6$ K, the star will radiate mostly in soft X-rays rather than in the visual range. Fortunately, the discovery of pulsars in 1968 has come to his rescue, for it is now clear that the pulsars are none other than rotating magnetized neutron stars. We therefore turn our attention to this interesting aspect of neutron stars.

Pulsars

The discovery of the first pulsar was reported in 1968. By now several hundred pulsars are known. Some of the striking features seen in the first pulsar are seen to be common to all of them. These features are summarized below.

First, the pulse period is very short (of the order of 1 s) for an astronomical object. The shortest known period is 0.033 s (second) for the pulsar PSR 0531 + 21 found in the Crab Nebula. Next, the period is remarkably steady and is known with a large accuracy. For example, the period of the pulsar PSR 0301 + 19 is known to an accuracy of 0.02 nanoseconds (i.e. 2 parts in 10^{11}). The steadiness and constancy of the period, however, refers to short term changes. Over a long period, most pulsar periods show a secular increase of the order of a few parts in 10^{15}. Since the periods themselves are of the order of 1 s, for a substantial change in the period at the above rate we have to wait for 10^7–10^8 years. This is the characteristic scale of the life-time of a pulsar. In addition a few pulsars show 'glitches', i.e. sudden decreases in period which subsequently increase again in the above secular fashion.

Although a few pulsars show emission in the optical and X-rays, by and large pulsars belong to radio astronomy through which they were first discovered. When the pulses from a typical pulsar are analysed they show individuality and a microstructure on the scale of $\sim 10^{-6}$ s. We will return to this fine scale aspect later. The interesting feature that emerges from the analyses of several pulses is the so-called characteristic '*pulse window*'. This is obtained by adding together several hundred pulse shapes. This average pulse shape seems to be fixed for each pulsar and may be considered as its signature. Thus, just as fingerprints are used to identify criminals, so from a pulse window it would be possible to identify the pulsar (provided of course, like the criminal, the pulsar is a known one with a previous record!).

When pulsars were first discovered several theories were put forward to account for their radio emission. Some of the ideas include compact close binary stars, pulsating neutron stars or white dwarfs, rotating neutron stars, etc. The reason why all these ideas revolved round very dense stars is not difficult to see. If ρ is the characteristic density and G the gravitational constant, then the time-scale of pulsation that emerges from all these ideas is of the order of

$$\frac{1}{\sqrt{(G\rho)}}.$$

Setting this equal to 1 s, we get $\rho > 10^7$ times the density of water. These densities are characteristic of matter in white dwarfs or the outer layers of neutron stars. The observational constraints, however, gradually narrowed down the choice of theories and all of them now centre on a rotating neutron star. Thus pulsars can be considered evidence for the existence of neutron stars.

A neutron star acquires rotational energy from the store of gravitational energy of the collapsing core and it may be rotating several times a second. The gravitational binding of a neutron star is strong enough to hold it together at such a rapid rate of rotation. A less dense star, even a white dwarf, would fly apart if it were to rotate so fast. Rotation alone is not enough for generating the observed pulses. The presence of a magnetic field is also demanded by pulsar theories. Just as in the Earth the rotation axis is aligned differently (i.e. at an angle) to the magnetic north–south axis, so in a pulsar the two axes are inclined. In some cases the angle between the two may be as large as 90°, implying that the magnetic poles lie on the equator of the star! The neutron star builds up a high density of matter by contraction: so it also builds up a very strong magnetic field near its surface. The field may be as high as 10^{12}–10^{13} Tesla, i.e. around a *million million* times the magnetic field on the surface of the Earth.

The theory first put forward by T. Gold in 1968 involves radio emission from a pulsar atmosphere of electrically charged particles. In spite of the very strong surface gravity of the pulsar (see Chapter 2) these particles can 'float' above the surface because the strong pulsar magnetic field is able to lift them. This atmosphere is carried along by the pulsar in the direction of its rotation. As we move further and further away from the rotation axis, the speed of charged particles increases, much in the same way as the outer parts of a merry-go-round move faster than inner ones. So at a large enough distance the electric charges move at speeds close to the speed of light. It is such particles, moving in circular orbits at near-light speeds, that emit radio waves by the process described in Appendix E. An alternative theory by V. Radhakrishnan and D. J. Cooke, however, suggests that the emission takes place near the surface of the pulsar at its magnetic poles. In such a theory it is necessary to describe how the radiation propagates through a pulsar's electrically charged atmosphere. In either case the radiation is beamed in a particular direction which keeps changing like the beam from the revolving lamp of a lighthouse.

The rotation of the pulsar tends to slow down because of magnetic braking. The magnetic lines of force tend to be pulled round by the rotating neutron star and they resist being twisted too much. This is the cause of the observed increase in pulsar period. If the pulsar starts life as a rapidly rotating object, then it will cease to exist as a pulsar when its rotation is substantially lost. The overall 'life-time' for this to happen is of course determined by the magnetic field in the pulsar, but is in the range of $\sim 10^7$–10^8 years for viable models.

If we consider the number of observed pulsars in the nearby region of our Sun, say out to a distance of 1500 light years, and use this number to estimate the *total* number of pulsars in the Galaxy, this latter figure comes out as high as 2×10^5. We do not expect to observe all these pulsars, partly because we could not see the more distant ones and partly because the pulsar radiation is beamed in a narrow region of space. If there were no reduction due to the beaming effect, the life-time and total number given above mean that there are on average two to three pulsars born per century – roughly at the same rate that supernovae explode in the Galaxy.

This is consistent with our picture of a pulsar being a neutron star born out of a supernova. There is also observational support to the extent that a pulsar (PSR 0531 + 21) has been located in the Crab Nebula, while another pulsar (PSR 0833–45) is seen near the Vela supernova remnant. The presence of a pulsar in the Crab Nebula also helps explain another mysterious feature of this remarkable object. The Crab is an emitter of radio waves, X-rays, and γ-rays as well as cosmic rays and it has been a problem to understand what energy machine keeps the Crab going. The supernova explosion observed in 1054 A.D. would hardly be expected to supply power to generate the activity in the Crab nine centuries later. The pulsar in the Crab Nebula, however, can supply the power for this activity from its vast store of gravitational energy. Several mechanisms including the synchrotron process described in Appendix E have been suggested to show how this can be achieved. We will not go into these theories here. We simply note the following result. If we estimate how much energy the neutron star inside the Crab is losing as its rotation slows down, the answer comes out to be at the rate of 3×10^{31} watts. This figure is more than ten times the rate at which radiation energy is observed to be coming out of the nebula in radio, visual, X-rays, etc. – i.e. in all forms of electromagnetic spectrum. Thus the balance may well be made up by cosmic rays.

Nevertheless the link between pulsars and supernovae need not be considered firmly established. Of the several hundred pulsars so far discovered only the *two* mentioned above are directly seen to be associated with supernova remnants. To the empirical astronomer relying only on firm observational evidence, the case is therefore far from made. One has to argue that the supernova exploded in an asymmetric fashion leading to a relative speed between the pulsar and the ejected envelope, much like the recoil suffered by a gun after firing a bullet. This may explain why the pulsar is not found within the supernova remnant.

Finally we refer to the glitches indicating a sudden decrease in the pulsar period. How can a rotating body suddenly increase its angular speed? A dancer figure-skating on ice changes his spin by throwing his hands out or drawing them in. The former process increases the dancer's moment of inertia and decreases his angular speed while the latter process produces the reverse effect. If in the pulsar a redistribution of matter takes place on its surface so as to *lower* its moment of inertia, its angular speed will increase and its period will decrease. To produce a glitch, this re-distribution must occur suddenly and it has been suggested that a *starquake* (like an earthquake on the Earth!) would achieve the effect.

In any case the microstructure in the pulse can tell us, in principle, minute details about the pulsar. A rule of thumb is that a pulse feature of time-scale t corresponds to a space feature of size ct in the pulsar. A time-scale of 10^{-6} s therefore corresponds to a size of 300 metres. The starquakes lasting a third of a nanosecond would tell us about surface discontinuities of the order of 10 centimetres! That such small scale information is available about the pulsar is a tribute to its highly accurate and stable time-keeping.

Gravitational Radiation

The accurate time-keeping in a pulsar has enabled astronomers to verify some of the subtle effects of general relativity. The pulsar PSR 1913 + 16 which has made this possible is somewhat exceptional in that it is part of a binary system. The pulsar was first discovered in July 1974 by R. A. Hulse and J. H. Taylor and it showed an orbital Doppler effect. This means that as the pulsar goes round its companion star of the binary system it alternately approaches and recedes from us. This results in the Doppler blue-shifts and red-shifts of its radiation.

The interesting result which emerged in 1979, after a detailed radio monitoring of this pulsar with the help of the 1000 foot dish at Arecibo, Puerto Rico, was that the orbital period of the system was decreasing at the rate $\sim 3 \times 10^{-12}$. (The orbital period itself was measured at 27 907 s, i.e. about $7\frac{3}{4}$ hours.)

Why should the orbital period *decrease*? If we assume general relativity to hold we have an explanation in terms of *gravity waves*. Recall that while discussing the defects of the Newtonian law of gravitation we mentioned that this law was inconsistent with special relativity because it implied that gravitational effects propagate instantaneously. This defect was corrected by general relativity in which the gravitational effects are known to propagate with the speed of light. The propagation of gravitational effects in general relativity is in fact a considerably involved problem when space-time geometry is highly non-Euclidean. The above remark that gravitational effects propagate with the speed of light refers to situations of weak gravity, when the space-time geometry is nearly Euclidean. Under such conditions there is considerable similarity between the phenomenon of electromagnetic radiation and the phenomenon of gravitational radiation.

The simplest source of electromagnetic radiation is an oscillating electric charge. The to and fro motion of the electric charge generates electromagnetic disturbances which travel away from the charge in the form of waves with the speed of light. The waves carry the energy of the moving charge. This loss of energy results in a damping of the motion of the electric charge so that the amplitude of its oscillation steadily decreases, in much the same way as the amplitude of the bob of a simple pendulum decreases due to the resistance of the air in which it oscillates.

The binary star system is similarly one of the simplest sources of gravitational radiation. As the two stars move round each other their gravitational effects in any region of space change with time. These effects travel outwards from the system carrying energy in the form of gravity waves. This energy comes out of the motion of stars which as a result gradually move along smaller orbits with smaller periods. This is what may be happenning in the case of the binary system associated with PSR 1913 + 16.

Taylor, Fowler, and McCullogh, who had been monitoring the pulsar, not only noted the decrease in the orbital period of the system but were also able to measure the other parameters of the binary orbits. This enabled them to quantitatively estimate the energy loss through gravitational radiation and the expected rate of decrease of the orbital period. This latter estimate agrees with the observed value very well.

While this observation does not conclusively establish the existence of gravity waves, it at least makes a good case for them. In any case no other satisfactory explanation for period decrease has yet been advanced. If the interpretation given by Taylor and his collaborators holds, then this is the first indirect detection of gravitational radiation. Those attempting to detect gravity waves in the laboratory have not yet achieved the sensitivity to be able to detect gravity waves from binary stars directly. How weak the gravity waves are and how difficult it is to detect them directly is discussed in Box 5.2.

Box 5.2 The sources and detectors of gravity waves

The basic source of gravitational waves in the 'weak-field' version of general relativity is a massive system whose quadrupole moment changes non-uniformly with time. More precisely the result may be stated thus. Using rectangular Cartesian coordinates (x_1, x_2, x_3) with $r^2 = x_1^2 + x_2^2 + x_3^2$ and specifying the matter density of the system by ρ, we define the quadrupole moment tensor by

$$I_{jk} = \int \rho(x_j x_k - \tfrac{1}{3}\delta_{jk} r^2) \, d^3 x.$$

Then the total power radiated in the form of gravity waves is

$$L = \sum_{j=1}^{3} \sum_{K=1}^{3} \frac{G}{45c^5} \dddot{I}_{jk} \dddot{I}_{jk} \quad (\cdot \text{ denotes rate of change}).$$

In practice L has to be averaged over several characteristic periods of the source. We will denote this average by $\langle L \rangle$.

$\langle L \rangle$ is very small for a laboratory system. To give an example: for a massive steel beam of radius 1 m, length 20 m, density 7800 kg/m^3, i.e. a mass of around 490 tonnes, the average power radiated by making the beam rotate about the axis through its middle cross-section and perpendicular to its length at a rate of about 28 radians per second is

$$\langle L \rangle \sim 10^{-29} \text{ watts}.$$

(At a higher rotation speed the beam may give way because of its limited tensile strength assumed here at 3×10^9 dyne/cm^2.) Clearly a laboratory source of gravity waves is out of the question!

Astrophysics does provide stronger sources of gravity waves. Binary star systems are ideal sources, although compared to electromagnetic sources these are still weak. The Jupiter–Sun system (Jupiter is massive enough to be taken as a 'star' here!) has $\langle L \rangle \sim 5200$ watts. An eclipsing binary like βLyr may radiate at a rate of $\sim 5.7 \times 10^{21}$ watts. Close binaries made of neutron stars or black holes could radiate at as high a rate as $\sim 10^{44} - 10^{49}$ watts.

Supernovae are also likely sources of *bursts* of gravity waves, provided sufficient distortions from spherical symmetry exist. In a characteristic burst the power output may be as high as $\sim 10^{47}$ watts and it may last for a fraction of a second. Accretion on to a black hole may also produce gravity waves, as also would a collision of black holes.

The direct detection of gravity waves in a terrestrial laboratory still awaits a technological breakthrough. Basically a gravity wave passing through space generates transient curvature in it which has to be detected by its effect on the motion of mechanical systems. The trouble lies in devising sensitive enough equipment to be able to detect these small effects of motion.

In the late 1960s, J. Weber from the University of Maryland first claimed that his apparatus had detected gravity waves from the direction of the centre of the Galaxy. Other experimentalists have, however, failed to detect gravity waves. Considerable controversy exists in this field with the majority view being that gravity waves have not yet been detected.

The binary pulsar shows two other effects peculiar to general relativity. One is the precession of the perihelion which we encountered for the planet Mercury. The precession seen in the binary orbit of PSR 1913 + 16 is 4.23 degrees per year – 36 000 times the effect in Mercury's orbit! The possibility that the precession could be caused by the tidal effects of one star on the other can be ruled out because both stars are likely to be compact neutron stars for which the tidal effect is negligible. Assuming that the tidal effect is negligible, the observed effect can in fact be explained entirely in terms of general relativity if the mass estimates of the two stars, at around $1.4 M_\odot$ each, turn out to be correct.

The second effect is associated with the precession of the rotation axis of the pulsar through what is known as the spin–orbit coupling. Spin–orbit coupling is a force which acts on a spinning body as it moves in a closed orbit around another body. The force disturbs the spinning direction (which would otherwise have remained fixed in space), causing it to precess. If the spinning body is a pulsar, its pulse pattern would also change as a result of this precession. This effect results in an estimated precession rate of $\sim 1°$ per year and would show up through slow changes in the pulse-window shape of the binary pulsar. There are indications that this effect is also present in PSR 1913 + 16.

Conclusion

This completes our discussion of stellar explosions and their after-effects. Both the envelope which is ejected and the core which survives in a supernova explosion have interesting effects like triggering star formation, contaminating the interstellar medium with material processed in the star, producing neutron stars and probably black holes, creating pulsars and cosmic rays, etc. Many aspects of the interaction of the supernova with its surroundings still remain to be explored and will no doubt receive important inputs from both theory and observations.

6 Powerful X-ray Sources in our Galaxy

Devouring all the worlds on every side with Thy flaming mouths, thou lickest them up. Thy fiery rays fill this whole Universe and scorch it with their fierce radiance, O Vishnu!

The Bhagavadgita XI-30

X-ray astronomy began in a very modest way soon after the Second World War, when small rockets became available for making observations from above ground-level. Because the Earth's atmosphere absorbs X-rays from outer space it is essential for X-ray astronomy that the X-ray detecting instruments are kept well above these absorbing layers. In fact the early observations of 1948 were motivated by reasons other than the purely astronomical ones. It had been noticed that daytime long range communication by radio waves occasionally broke down due to a 'fade-out'. The cause had been the appearance of free electrons in the Earth's atmospheric D-layer which began at a height of around 80 km. It was also noticed that the fade-out occurred at the time of solar flares. Could this phenomenon be explained by assuming that the flare generated X-rays which were capable of ionizing the atoms in the D-layer and releasing the electrons? The U.S. Naval Research Laboratory launched the rockets to find an answer to this question. The answer, in 1948, was affirmative, although not decisively so.

In 1956 T. Chubb and H. Friedman were able to obtain a conclusive proof by a series of rocket experiments. By good luck a large flare did break out on the Sun on the fourth day of their daily rocket launches and Chubb and Friedman were able to record an intense burst of X-rays. However, they also made another discovery through these experiments which was potentially more significant: they found X-rays coming from all over space in a diffused fashion (i.e. not from a few clear-cut sources).

The next discovery in X-ray astronomy also came unexpectedly. On 12 June 1962 a rocket was launched to a height of about 230 km by two groups of X-ray astronomers, B. Rossi and G. W. Clark from the Massachusetts Institute of Technology and R. Giacconi, F. Paolini, and H. Gursky of American Science and Engineering Inc. Although the aim of the experiment was to detect X-rays from the Sun scattered by the Moon, the X-ray detector also found a significant signal from one particular region of the sky, indicating a possible cosmic source of X-rays. As X-ray detection technology improved, it became possible to pinpoint more accurately the location of this source and by mid-1966 astronomers were able to identify the source, now known as Sco X-1 (*Sco*: short for the direction of the source in *Scorpius*), with a 13th magnitude* star.

As mentioned in Chapter 1, the UHURU satellite was launched in 1970. From 1970 onwards the progress of X-ray astronomy has been rapid and spectacular. The use of satellites, improved and more sensitive detectors, more accurate position locating equipment, and X-ray spectrum measuring instruments has progressively improved the information supplied by X-ray astronomy. The most important of these instruments has been the High Energy Astronomical Observatory-B (HEAO-B), launched in 1978 and renamed as the Einstein Observatory to commemorate the centenary of the birth of Albert Einstein. Rather than follow our historical approach any further, we shall now take stock of this improved information.

Types of X-ray Sources

By now X-ray emission has been found to occur in several types of astronomical objects. Soon after Sco X-1, Friedman and his colleagues found in 1963 X-ray emission from the Crab Nebula. These early observations were, however, not precise enough to pinpoint the location of the source in the nebula. Did the X-rays come from the diffuse nebulosity or did they come from a point source somewhere inside the nebula?

In 1964 this issue was resolved by the fortunate circumstance of the Moon happening to cross the path of the nebula. If the Moon blocked the radiation sharply, this would imply the existence of a point source. If on the other hand the occultation by the Moon produced a smooth decline in the X-ray emission (as in the case of a Solar eclipse) then the radiation must be from an extended region. Analysis of the 1964 observations indicated that at least 90 per cent of the X-rays came from the extended region.

The Crab Nebula, as we saw in the last chapter, has also the remarkable feature that it houses a pulsar. The Crab pulsar not only sends out radio pulses but it also emits optical pulses, at the same period of 0.033 s. In 1969 the X-ray observations revealed further the even more interesting result that the Crab pulsar was also emitting X-ray pulses. In fact, since the typical X-ray photon is much more energetic than the optical or the radio photon, if we calculate the total amount of energy emitted by the nebula in

---

* Magnitude is a measure of a star's brightness on a logarithmic scale. See *Glossary* for details.

these different forms, we find that the energy emitted in the X-rays is the largest (about five times that in the radio region).

It is very likely that the Crab pulsar is powered by the rotating magnetized neutron star in the manner discussed in the last chapter and that X-rays from it arise in this way. It is not clear if other supernova remnants are X-ray emitters like the Crab, powered in the same way by their central pulsars. Some of the other X-ray sources (besides the Crab) which are associated with supernova remnants are listed in Table 6.1 below.

There are other X-ray sources in the Galaxy, besides the supernova remnants. An important class of sources is found amongst close binary systems. The source Sco X-1 has been listed as a single star in Table 6.1. At the time of writing it has not been definitely established as a binary source, although there is evidence of periodicity which could be indicative of a binary period. We will discuss the binary X-ray sources in the next section.

The binary sources are by and large *variable* in their X-ray output. However, another type of X-radiation was observed in 1975 from several sources in the Galaxy: this radiation was in short bursts. The bursts last for several tens of seconds and may be repeated at regular intervals of the order of hours or even days. One particular burst source—known as the *Rapid Burster*—is exceptional in the sense that it emits bursts like a rapid fire machine gun, sending out several thousand bursts per day. In a typical 10-second burst the X-ray flux emitted may exceed the Sun's total energy flux (at all wavelengths of the electromagnetic radiation) for one week! We will explore later the possible energy machines which may be powering such burst sources.

Finally, X-ray sources exist on a much larger scale than those found in the Galaxy. Other galaxies beyond ours, and quasars, are powerful emitters of X-rays. Since we intend to discuss these extragalactic objects in the next chapter, we will postpone the description of their X-ray emission to that chapter. Here we simply list some of them in the following table of typical X-ray sources.

Table 6.1 below describes the sources by their UHURU catalogue number (3U denotes the 3rd UHURU catalogue). As mentioned in Chapter 1, the UHURU satellite provided the first comprehensive list of galactic and extragalactic sources. The list below is a partial one intended to convey a flavour of the variety of sources.

Binary X-ray Sources

The sources Hercules X-1 and Cygnus X-1 listed in Table 6.1 occur in close binary systems. The connection of X-ray sources with binary systems was first established for the source Her X-1 when this source was

Table 6.1 A Few Typical X-ray Sources

	Source type and examples	*3U-catalogue number*	*Approximate distance in light years*	*Approximate X-ray power in watts in 2–10 keV*	*Remarks*
GALACTIC SOURCES	A. *Supernova remnants*				
	Tycho's Nova	0022+63	16×10^3	5×10^{28}	Supernova of 1572
	Crab	0531+21	6.5×10^3	10^{30}	Supernova of 1054 Also a pulsar
	Vela X	0833−45	1.6×10^3	10^{27}	
	Cassiopeia A	2321+58	11×10^3	10^{29}	
	B. *Close binaries*				
	Hercules X-1	1653+35	$(6\text{–}20) \times 10^3$	10^{29}–10^{30}	1.7 day eclipsing binary
	Cygnus X-1	1956+35	6.5×10^3	5×10^{29}	5.6 day optical period
	C. *X-ray stars*				
	Scorpius X-1	1617−15	1.6×10^3	5×10^{29}	Optical counterpart V 818 Sco
	D. *Globular clusters*				
	NGC 6624	1820−30	20×10^3	10^{30}	X-ray burster
EXTRAGALACTIC SOURCES	E. *Galaxies*				
	M 31	0021+42	2×10^6	10^{32}	Spiral galaxy
	NGC 5128	1322−42	2×10^7	5×10^{34}	Radio source Centaurus A
	F. *Quasars*				
	3C-273	1224+02	3×10^9	5×10^{38}	Quasar of red-shift 0.158
	G. *Cluster of galaxies*				
	Virgo	1228+12	4.4×10^7	7×10^{35}	M87 is the main source

identified with the star HZ Hercules. How the identification was established can be briefly seen as follows.

Her X-1 pulsates with a time-scale of 1.24 seconds. Like the Crab pulsar, this source also houses an X-ray pulsar. However, unlike the Crab, the

X-rays from this source undergo 'eclipses' every 1.7 days. Also the source becomes unobservable every $\sim$ 35 days for 20 days or more. How does a binary star model explain these time-scales and fade-outs?

In Fig. 6.1 we see a close binary star model. In such a system we have two stars A and B going round each other, with their surfaces very *close* to each other as shown in the figure. If we observe such a system from the *plane* of the orbit we would see eclipses when one member of the binary gets into our line of sight to the other member. Box 6.1 shows how to relate the

---

Box 6.1 The dynamics of a binary system

We refer to Fig. 6.1 of a typical binary system consisting of stars A and B going round each other. What in fact happens is that the point C situated at the position of the centre of mass of A and B remains fixed (or at least unaccelerated) in space. Let m_A and m_B be the masses of A and B. Then we have

$$m_A \times AC = m_B \times BC.$$

Let us suppose, in a simplified picture, that the orbits are nearly circular, so that $AC = a = $ constant and $BC = b = $ constant, throughout. If P is the common period of the orbits then we have the orbital velocities v_A and v_B of the stars A and B given by

$$P = \frac{2\pi a}{v_A} = \frac{2\pi b}{v_B}.$$

Therefore we have

$$\frac{v_A}{v_B} = \frac{a}{b} = \frac{m_B}{m_A}.$$

By observing the way the spectral lines of the stars move backwards and forwards across their spectra due to the Doppler effect, it is usually possible to measure the ratio of the orbital velocities $v_A : v_B$. The period P is also determined from this observation since one backward and forward shift of lines is completed in this period. If we knew the inclination of the plane of the orbits to the line of sight then the spectroscopic observations could lead to the measurements of v_A and v_B *separately*. Otherwise, as is usual, only their ratio is known.

Knowing P and $a + b$ we can determine $m_A + m_B$ by Kepler's third law of motion (see Box 2.2). The answer is

$$m_A + m_B = \frac{4\pi^2 (a + b)^3}{P^2 G}.$$

Thus if v_A, v_B are *separately* known we can estimate a and b from the period P. Then the above formula gives $m_A + m_B$. We also know $m_A : m_B$ as equal to the ratio $v_B : v_A$. Thus we know m_A and m_B individually. However, if we do not know the inclination of the binary orbit we can only place lower limits on m_A and m_B.

In the case of Cygnus X-1, the supergiant star (A) only is seen. Its spectral lines show a periodicity of $P = 5.61$ days. *If* we assume the orbital plane to contain the direction of line of sight, we can estimate v_A from the Doppler effect on the spectrum of A. This assumption enables us to place a *lower* limit on the mass of the other (unseen) star B at $\sim 5\ M_\odot$.

time-scale for these eclipses to the orbital parameters of the system. For observers like us the binary system is then known as an *eclipsing binary*.

In the case of Her X-1, there was an optically variable star with a period of 1.7 days, the star HZ Hercules mentioned before. This is an F-type star whose light output drops with the above period—a phenomenon which could be explained if it were part of an eclipsing binary. The coincidence of the 1.7 day period in the X-ray eclipses of Her X-1 and in the optical variations of HZ Hercules therefore strongly suggests their being part of the same system. The positional accuracy of X-ray measurements is good enough for the observers to assert that the X-ray source and the optical source must be part of the same system.

What about the companion star of HZ Hercules? The fact that it is not 'seen' by the optical astronomer, as well as the observation that the light of HZ Hercules is not totally cut off by its companion, suggests the companion to be a compact star(a star of normal size or a giant star would have been easier to spot and it would have eclipsed HZ Hercules more effectively than a small size star can). The calculations of the orbital parameters as given in Box 6.1 place the estimate of the mass of the compact companion at around $1\ M_\odot$. This mass limit lies well within the upper limit for a stable neutron star. We therefore have a situation where Her X-1 is a close binary system with one star a giant star and the other a neutron star.

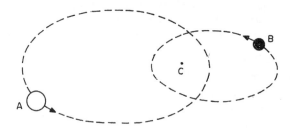

Fig. 6.1. The two stars A and B in a binary system go round each other in elliptical orbits such that their common centre of mass C is fixed. In a close binary system the separation of the centres of the stars exceeds the sum of their radii by a small fraction.

This scenario is found to be a common one for binary X-ray sources, and before considering how the X-rays are generated it is interesting to look at how the binary star system itself evolves to this stage. Figure 6.2 illustrates the typical evolutionary sequence in four stages. We start in stage (i) with a pair of stars A and B with masses 8 $M_\odot$ and 20 $M_\odot$ respectively. Star B, being more massive, evolves more rapidly. After 6.2 million years, B becomes a giant star and acquires a radius so large that it cannot hold itself together under the tidal forces exerted by its companion A. As a result the matter from B begins to flow towards A in the manner shown in stage (ii). The dotted horizontal figure of eight shown in (i) and (ii) is the so-called *Roche lobe* (named after E. Roche who first pointed out

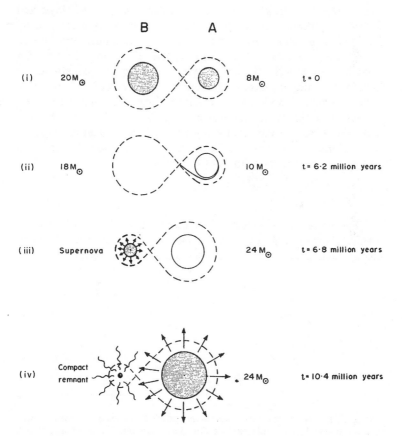

Fig. 6.2. Four stages in the evolution of a binary star system. (Based on computations by C. de Loore and J.-P. de Grève.)

in 1850 the possibility of tidal disruption exerted by a planet on its satellite). This lobe decides the maximum extents for the two stars for them to be able to retain undisrupted shapes. After 6.8 million years, star B explodes as a supernova (see Chapter 5) leaving behind a neutron star. The mass of star A, meanwhile, has grown to 24 $M_\odot$ as a result of accretion of matter from star B. This is shown in stage (iii). Finally stage (iv) brings us to the characteristic situation of a binary X-ray source. In this situation the star A has become a supergiant and matter from *its* surface crosses the Roche lobe and begins to flow to star B. This 'stellar wind' is the cause of X-radiation in the manner to be discussed now.

Accretion Discs

In the 1940s three British astronomers, Hermann Bondi, Fred Hoyle, and Raymond Lyttleton, had considered situations where a star accretes matter from the surrounding medium because of its gravitational attraction. In Fig. 6.3 we see a spherical star S in motion relative to the interstellar medium. If we consider the star as being at rest, the distant matter on its right (in the figure) appears to move towards it with a speed v. Provided the initial direction of motion of the distant particle in the interstellar space passes close to the star, the star will exert significant pull on it and bend its path. The thick lines indicate the boundary of the cylinder in which the particle's initial trajectory must lie in order that it is affected appreciably by the star's gravity. The radius of this cylinder is

$$R_g = \frac{4GM}{cv}.$$

When matter in the cylinder comes near the star its path is bent in the manner shown in Fig. 6.3. Such particles coming from many directions collide and lose their transverse momentum and eventually fall on the star.

Fig. 6.3. The spherical star S makes a distant particle P, approaching it with velocity v, turn round and go in a different direction as shown above. If the line of approach of P lies a small enough distance from the centre of the star the particle is eventually captured by S. The thick lines show the boundaries of the cylinder within which P must approach S for capture and accretion.

If the density of interstellar particles is ρ, the rate of accretion (that is, the rate at which the star picks up mass from the ambient space) is given by

$$\dot{M} = \pi R_g^2 v \rho = \frac{16\pi G^2 M^2}{c^2 v}.$$

What relevance has all this to the binary X-ray sources? The relevance is seen from the fact that in a typical X-ray binary the compact star accretes matter from its companion. The basic idea is the same here as that discussed by Bondi, Hoyle, and Lyttleton for stellar accretion. The details differ in the following respects.

First, the rate calculated above is for accretion of dust particles. In 1952 Bondi showed that the rate could be significantly stepped up provided we assume the accreted matter to be in the form of a fluid. The pressure and the collisions in the fluid play an important role in that they reduce the transverse momentum of the incoming matter and augment its radial momentum so that the rate of accretion is stepped up. In the binary X-ray source the accreted matter, that is, the gas from the companion star's surface, behaves like a fluid.

The second important difference in the scenario which steps up the accretion rate is the intense mass loss from the companion star close to the compact star. Instead of drawing matter from a diffuse interstellar medium, the compact star has a nearby source from which to capture matter.

The third and the most important point of difference in detail is introduced by rotation. In general, not only do the two stars in the binary system go round each other, but they also rotate like the Earth about its north–south axis. Rotation affects the trajectories of the accreted matter in a manner illustrated in Fig. 6.4.

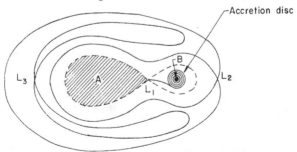

'Fig. 6.4. The accretion disc is formed by matter which escapes from the surface of star A through the Lagrangian point L_1 and falls into star B. The Lagrangian points L_1, L_2, and L_3 have special significance in the dynamics of particles moving under the gravitational influence of A and B in the rotating binary system. The continuous lines indicate equipotential surfaces.

As shown in Fig. 6.4, the matter from the supergiant companion escapes through the point L_1, where the Roche lobe intersects itself. The escaped matter goes round the compact star. It may either leave the system altogether through the vicinity of point L_2, or it may come back to star A through L_1, or it may circulate round the compact star B. It is the last possibility that we are concerned with here. This matter after circulating for a while loses its angular momentum through collisions and gains energy as it falls into B, attracted by its strong gravity. Since the process goes on continuously, the circulating matter forms a disc round B, a disc which is usually known as the *'accretion disc'*.

Some mathematical details of the accretion disc are given in Box 6.2. Qualitatively we see that the disc acts as a source of radiation whose

---

Box 6.2 Accretion discs

In the last few years considerable work has gone into determining the structure of accretion discs, with attempts to find models which would remain stable. We do not wish to enter into those details here. If a stable disc does form, it radiates like a black body with its spectrum peaking at photons of energy

$$E = \left[\frac{\dot{M}_0}{10^{-9} \, M_\odot/\text{year}} \right]^{\frac{1}{4}} \left(\frac{M}{M_\odot} \right)^{-\frac{1}{2}} \text{keV}.$$

For generation of soft X-rays, E should be in the range of around 1–10 keV. For a neutron star or a black hole of mass a few times $M_\odot$ we see that the above formula gives emission in soft X-rays for an accretion rate $\dot{M}_0 \sim 10^{-9} \, M_\odot$/year. This accretion rate is considered reasonable in a close binary system.

A critical role is played in the accretion process by the *Eddington luminosity* which is given by

$$L_c = \frac{4\pi G M m_p c}{\sigma}$$

where M = mass of the compact star B and m_p is the mass of a proton. The quantity in the denominator is the scattering cross-section of light by an electron. What does L_c imply? If the luminosity of the X-rays generated by the disc exceeds L_c, the photons are able to disrupt the gravitationally bound disc. If the luminosity falls significantly below L_c, the gravitational binding wins. Putting in numbers we get

$$L_c \cong 5 \times 10^{31} \left(\frac{M}{M_\odot} \right) \text{watts}.$$

For M a few times $M_\odot$ we get $L_c \sim 10^{32}$ watts. This is in the region of the luminosity of typical binary X-ray sources.

energy source primarily lies in the strong gravitational pull of the compact star. It is because of the star's gravity that the material falls into it with increasing speed. The viscosity of the material as it falls in causes it to heat up. Viscosity is a form of friction to which fluids are subjected whenever the neighbouring layers of the fluid move relative to one another – like the frictional force which tries to prevent the sliding of one solid surface over another. In the case of the accretion disc, viscosity arises because differences of velocity exist as we progressively go inwards. The viscous heating takes away the kinetic energy of the infalling material. For the typical compact star mass $\sim M_\odot$ the temperature of the disc rises to $\sim 10^7$ K and a body of such high temperatures largely emits soft X-rays of the type observed from typical X-ray binaries.

Considerable theoretical work has gone into the structural properties of accretion discs since their importance for X-ray binaries was realized. The early models were relatively simple and these turned out to be unstable! That is, the disc, even if it were formed, would be blown apart by instabilities. As a result, more complicated models were constructed which satisfied the established criteria of stability. Although there is still no unique disc model accepted by all astrophysicists, it is by and large assumed that disc radiation is the best possible explanation of X-radiation from binaries.

For neutron stars there is another way of generating X-rays. In Chapter 5 we saw that neutron stars may have strong magnetic fields, although their magnetic axes may not be aligned with the rotation axis. In a magnetized neutron star in a binary, the ambient charged matter may preferentially fall along the polar axis on to the poles of the star much like liquid being poured down a funnel. In such a case the infalling matter radiates X-rays and the polar regions would correspond to the 'hot spots' of the neutron star. The corresponding X-rays come along the polar axis.

How do we account for the on-off property of X-rays from sources like Her X-1? There are several possibilities which may cause the observed interruptions. The periodicity of interruptions suggests the precession of the disc as the best possible explanation. Just as a spinning top wobbles about its axis, so the disc around the rotating compact star wobbles. And this wobble results in our being periodically cut off from the direction (along the polar hot spots) in which the X-rays are emitted by the neutron star.

Another effect which may play an important role in the emission of X-rays is the competition between the inward pull of gravity and the outward push of the emitted radiation. This is an effect which was discussed by Sir Arthur Eddington in the 1920s in connection with ordinary stars and which tells us that if the radiation from the disc exceeds a certain critical value its outward pressure will stop further accretion and will, in fact, blow the disc apart. On the other hand, if the luminosity falls well below the critical value, the accreted matter may swamp the radiation and cut it off! This critical value of luminosity is known as the *Eddington luminosity*. As shown in Box 6.2, for a star of mass a few times $M_\odot$ this luminosity is around 10^{32} watts. Most X-ray binaries do have luminosities of this order.

Does Cygnus X-1 House a Black Hole?

Perhaps the most interesting source among the X-ray binaries is Cygnus X-1, because of its possible connection with a black hole. This source has been identified with a binary system of which the visible member is a supergiant star HDE 226868. The period of the binary system, determined from its optical properties, is 5.6 days. The source does not show regular X-ray pulses like Her X-1, but it shows variations of X-ray luminosity of the order of $0.1 - 10$ s. Some fluctuations appear to last for as short a time as 0.001 s.

In 1971 a faint radio source was detected in the vicinity of Cygnus X-1 by L. Braes and G. Miley, and by C. Wade and R. Hjellming. The variations of radio flux coincided with the variations of X-ray flux, leading to the conclusion that the X-ray source and the radio source constitute one and the same object. In fact this fortunate circumstance helped in the identification of the X-ray source with the optical binary system. For, unlike the radio sources which are comparatively rare, the stars (including binaries) are very common so that, unless the position of the X-ray source is very accurately known, it is a difficult problem to identify the precise optical object associated with it. Although the present X-ray telescopes can pinpoint a source within an angle of a few arc seconds, the source Cygnus X-1 could only be located by UHURU in an angular area of 4 square minutes. On the other hand, the position of the radio source connected with Cygnus X-1 was known within an angular error of 1″. It thus became possible to single out HDE 226868 as the binary star generating the X-rays from Cygnus X-1. The more accurate X-ray detector in the Einstein Observatory has confirmed this identification.

The visible component of the binary system is a B-type star. In the spectral classification scheme of stars, such B-type stars are massive and luminous. From the general information available about masses of these stars, the mass of the star HDE 226868 is estimated to be *at least* 20 $M_\odot$. The period of the binary is 5.6 days. The velocity of the visible component in the radial direction can be estimated from its Doppler shift (see Appendix C). From consideration of Box 6.1 the mass of the invisible companion can be estimated to be *at least* 5 $M_\odot$. The reason for the qualification 'at least' is that we are not necessarily in the orbital plane of the binary system and also because our estimate of the mass of the extended companion is a *lower limit*. Hence we cannot estimate the mass of the compact object *exactly*; we can only estimate the limit that the mass must *exceed*.

Even this lower limit of 5 $M_\odot$ is, however, well above the limit of the mass of a neutron star discussed in Chapter 5. What can the compact object be, if it is not a neutron star? If its small time-scale fluctuations are anything to go by, the object must be extremely compact, even more so than a neutron star. The time-scale of fluctuations, 0.001 s, when translated into a distance scale by multiplication by the speed of light, gives a distance of 300 km. From relativity theory we know that no physical disturbance can travel faster than light. Hence any coherent physical process which produces fluctuations as rapid as a thousandth of a second cannot extend over a region larger than 300 km in size. The accretion disc has to be as small as this in size, if it is to generate such rapid fluctuations of X-rays. The high intensity of X-rays coming from Cygnus X-1 also enables theoreticians to conclude that the emitting source must be a highly compact object.

Evidence of this kind makes Cygnus X-1 unique amongst X-ray binaries. Since no other type of compact star can fill the bill, the conclusion usually drawn is that the unseen member of this binary system is a *black hole*. If this conclusion stands, then X-ray astronomy can claim to be the first to have led to the discovery of a black hole!

To the sceptic, however, the available evidence is not convincing enough. Can the black hole explain all the time-scales of fluctuations associated with Cygnus X-1? Can it not be the case that the compact object itself may be made up of two bits, say two neutron stars? Have we really understood the state of matter at densities beyond the limit of $\sim 10^{15}$ kg per dm^3 of neutron star matter, to be able to assert that *no* compact star can exist with a mass as high as 5 $M_\odot$?

The sceptics, of course, have to produce viable models either of 3 bodies moving in one another's gravity (instead of the 2 in a binary system), or of dense compact stars of masses exceeding 5 $M_\odot$, to combat the black hole lobby for Cygnus X-1.

Even if the sceptics fail to achieve this, the purist may still record his dissatisfaction with the black hole interpretation on one count. *None* of the observed features of *Cygnus X-1* confront us with the remarkable space-time effects described in Chapter 4 which are associated with the horizon and the ergosphere and which are peculiar to black holes only. To the purist, all observable aspects of X-rays from Cygnus X-1 convey direct information only about the accretion disc which lies well outside the horizon.

X-ray Bursters

We now leave the binary X-ray sources and turn to another type of galactic source discovered in 1975. More than 30 *X-ray bursters* are now known, most of them lying in the disc of the Galaxy, close towards the centre. The Rapid Burster referred to earlier lies in a globular cluster which was identified only after the X-ray source was discovered. Statistics to date are rather poor and it is not possible yet to say if there is a link between bursters and globular clusters. In any case, the discovery of bursters has regenerated interest in the study of globular clusters with the hope of understanding the mechanism of X-ray emission found amongst them.

It was William Herschel in the 1780s who first recognized the globular clusters for what they are: a collection of stars gravitationally bound. A typical globular cluster may contain as many as 100 000 stars. In Fig. 1.5 we see the globular cluster NGC 6624, which has been identified with the X-ray source whose UHURU catalogue number is 3U 1820–30. The centre of the cluster is thickly packed with stars whose individual images cannot be resolved in this photograph. The estimated star density in the central region is around 3000 $M_\odot$ per cubic light year—a star density a million times more than that found in the vicinity of the Sun. A crowded region, indeed! But then this is a common feature of all globular clusters.

Before concerning ourselves with the X-ray sources in globular clusters let us briefly review the current ideas on the origin and evolution of the clusters themselves. It is conjectured that our Galaxy as a whole was formed by the condensation of a gigantic primordial cloud made up largely of hydrogen and helium. During the condensation not all of the Galaxy was formed simultaneously. Rather, the inhomogeneities of the gas cloud resulted in the condensation of dense regions first. Such dense regions condensed into star clusters, each cluster containing around a *million* stars. A cluster then moved as a single unit through the protocloud. As it is believed that the cloud as a whole possesses angular motion about an axis, the clusters also move round the axis, completing one round in several million years. There is, however, relative motion between the cluster and the protocloud and this motion strips off all the gas in the cluster which has not already condensed into star-like units. Therefore only that matter survives in a cluster which has already condensed into stars and so these surviving stars differ in their birth dates by less than a few million years.

What happens to these surviving stars in a globular cluster? Stars which were considerably more massive than the Sun completed their evolution quickly and became supernovae as outlined in Chapter 5. Since this time-scale may be as short as a million years for massive stars, these stars will have exploded even before the cluster completes one revolution. The heavy nuclei processed in these stars and ejected in explosions contaminate the surrounding medium so that any star clusters which form later are made up from this contaminated matter (rather than from the purely primordial mixture of helium and hydrogen).

The less massive stars with masses down to $1 M_\odot$ have a less dramatic future. They do not explode but eventually settle down to become white dwarfs. The time-scale of their evolution is much longer than a million years. Stars of very low mass (say less than about $0.5 M_\odot$) are swept out of the cluster by tidal forces of the disc of the Galaxy. Or they may acquire sufficient velocity through gravitational collisions in the cluster to be able to leave the cluster altogether. So what is left in the cluster is the group of stars with masses mainly in the range of $0.4 M_\odot$ to $0.8 M_\odot$ (stars which are still burning their nuclear fuel) and the remnants of the massive stars which exploded long ago. The X-ray sources may be connected with these remnants.

Various scenarios have been considered by theoreticians to account for the X-ray emission from globular clusters. To begin with, we note that the X-ray emission from globular clusters is comparable to strong X-ray emission from stellar binaries considered earlier. Therefore one explanation relies on the binary star picture described in this chapter. However, in order to sustain this picture we need a compact star and a normal nuclear fuel burning star in orbit round each other. The latter type, we have just seen, must be extinct by now, *if it was originally part of the cluster.* Such stars must therefore have been acquired by the cluster by capture from outside. The process of capture is not straightforward since even in the dense central regions of globular clusters the frequency of close encounters among *three* stars (which may result in the capture of one by the other while the third star escapes with a large velocity) is rare. Theoreticians A. Fabian, J. Pringle, and M. J. Rees from the University of Cambridge have, however, suggested that, under certain circumstances, in an encounter between *two* stars the tidal dissipation of energy can also lead to capture. J. Hills from the University of Cambridge has also found from computer simulation that if a binary system of two low mass stars encounters a massive remnant (e.g. a neutron star or a black hole) the encounter often results in an exchange between one low mass star and the massive remnant. The two processes are illustrated in Figs. 6.5(a) and (b).

If these scenarios are correct, we should expect several close binaries in the central region of the globular cluster. The difference between these binaries and those which give rise to X-rays in the disc of the Galaxy is that in the present case the companion of the compact star is itself of low mass. Such a star, when it becomes a giant (instead of the larger supergiant discussed previously), supplies accretion matter to the compact star. However, the supply is at a modest rate and for a longer time since the

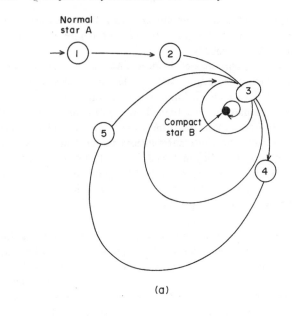

(a)

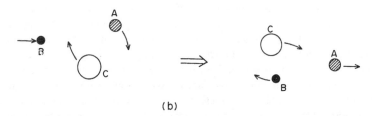

(b)

Fig. 6.5. The two processes whereby binary stars can form in the central region of a globular cluster by a two-body encounter. In (a) an ordinary nuclear fuel burning star A approaches the compact star B and its orbit round A gradually shrinks and becomes more and more circular as the system loses energy by tidal interaction. In (b) a compact star B exchanges place with a low mass star A in an already existing binary system (A, C). (Adapted from an article by G. W. Clark in *Scientific American*.)

expanded red giant phase of the supplier may last for hundreds of millions of years. The rate of accretion may show long-term variation as a result of the changes that occur in the giant star, especially late in its life. This may explain the long-term variability of the globular cluster X-ray sources.

A radically new hypothesis was put forward by J. P. Ostriker and J. N. Bahcall of Princeton and by J. Silk of California. This hypothesis is based on the assumption that the dense concentration of stars in the central region of a globular cluster results in the formation of a *massive black hole*. To date, it has not been possible to say how such a black hole with mass in the range of $100\,M_\odot$–$1000\,M_\odot$ may come about! We will nevertheless *assume* that such a black hole exists and explore its consequence for X-radiation.

A massive black hole attracts matter in its surroundings, mainly tenuous gas left behind in the globular cluster, and this infalling material picks up speed as it falls into the black hole. If the black hole is rotating this will preferentially generate an accretion disc of infalling material in its equatorial plane. The net result is the same as in the case of binary stars: the accretion disc will grow hotter and begin to radiate. The reason the black hole has to be a massive one (with a mass of about $100\,M_\odot$–$1000\,M_\odot$) is because the gas in the globular cluster tends to leak out and it needs a strong enough source of attraction to hold it in the centre.

What could be considered 'evidence' for such a massive black hole? The presence of hot luminous gas or even a sharp rise in the luminosity of the cluster towards the centre could be considered indicators of a strong attracting body in the centre. The rise in luminosity is caused by the dense concentration of stars attracted by the massive black hole.

The black hole idea did in the early days appear to receive some confirmatory evidence from the observations of X-ray *bursts* in the X-ray source 3U 1820–30 identified with the globular cluster NGC 6624 shown in Fig. 1.5. This source is one of the brightest UHURU sources in the central regions of the Galaxy. This object was observed during 1975 by the Astronomical Netherlands Satellite (ANS) and later by the Third Small Astronomy Satellite (SAS-3). The X-ray bursts seen in the energy range 1–30 keV were identified as coming from 3U 1820–30. The bursts lasted about 10 seconds, but the estimated energy in a single burst was of the order of 10^{32} J – about as much energy as the Sun would emit in a week! How could so much energy be released in such a short time?

E. Grindley and H. Gursky from the Harvard College Observatory suggested that the detailed structure of the burst consisting of a rapid rise time of $\sim 1\,\text{s}$ and a slow decay time of $\sim 10\,\text{s}$ implied that the X-ray photons emitted in the burst reverberated as they were scattered around. This reverberation is like that of sound waves in a temple after the temple bell has been tolled. But what causes the reverberation of X-rays in 3U 1820–30? Grindley and Gursky suggested that this happened when the X-rays were scattered by surrounding clouds of hot plasma. These authors were also able to explain with their model another feature of the burst phenomenon: the hardening of the spectrum. That is, they were able to explain how, as the total energy of the X-ray declined, the proportion of higher energy photons in it *increased*.

The hot gas or plasma in the Grindley–Gursky model could be held in the central region of NGC 6624 where the X-rays came from, provided there was a strong enough gravitational source – a source as massive as $1000 M_\odot$. Thus the massive black hole idea began to receive more adherents as this model gained more and more support. (C. R. Canezares has however a cold plasma model which requires a central mass less than $100 M_\odot$.)

Further examination of bursts from this source showed the following features besides the rise and decay times mentioned earlier. The bursts were repeated after ~ 4.4 hours. That is, their recovery time was $\sim 10^4$ s. But this was not constant: it varied by as much as ± 600 s. This variation, known as 'jitter', also required explanation. The proponents of the massive black hole idea, Bahcall and Ostriker, suggested that the burst and the jitter were caused by a neutron star going round the black hole. This star periodically crashed through the accretion disc and generated X-ray pulses with a stable average period. But as the disc itself flip-flopped randomly, there was a jitter in the period.

Recent observations from the Einstein Observatory do not seem to support the massive black hole model. Statistics do not bear out the contention of the model that X-ray sources are at the centres of globular clusters. An off-centre location goes against the massive black hole idea. Also recently Grindley has conceded theoretical difficulties with the hot plasma model based on a massive black hole.

However, another model *not* involving a massive black hole was proposed by F. Lamb from the University of Illinois. In Lamb's model the basic feature was as follows. A magnetized neutron star, like that described in Chapter 5, accretes plasma. The plasma may, however become trapped in the star's magnetosphere. It would begin to pile up at a certain height mainly above the polar regions of the star. However, beyond a certain critical density it can no longer be held trapped and it would spring a leak and crash down on the star. The crashing process converts the gravitational potential energy of the plasma first into kinetic energy and then into X-rays. After this has happened the 'hole' in the magnetosphere would be patched up and the plasma would again start piling up. This is how the burst could be explained. The jitter would be caused by certain other factors which determine the rate at which the plasma reservoir begins to fill up.

This latter model received support with the discovery of the so-called *Rapid Burster* which we will now describe in some detail.

The Rapid Burster

In 1976 W. Lewin and his colleagues detected, with the help of the SAS-3 satellite, a new type of X-ray burster. Labelled as MXB 1730 − 333, this source is remarkable in that it emits bursts of X-rays in a rapid-fire sequence, the interval between bursts in some cases being as small as 10 s. The source was first identified with a previously unknown compact cluster by W. Liller of the Harvard College Observatory who studied the star field in the vicinity of the source with the help of long-exposure red-sensitive plates taken with the 4 metre telescope at the Cerro Tololo Inter-American Observatory in Chile. Known as *Liller I*, this cluster was subsequently identified with a globular cluster by D. E. Kleinmann, S. G. Kleinmann, and E. L. Wright using infrared observations.

The Rapid Burster has another property which is very revealing. The recovery time after each burst is roughly proportional to the size of the burst, although the peak intensity hardly varies from burst to burst. (The *size* of the burst is the measure of the *total* energy in it and is given by the area under the intensity curve.) This property is like that of a neon lamp flasher. After each flash the oscillator in the lamp must recover for a length of time proportional to the depth of the discharge, which varies from flash to flash. As in the case of the neon flasher, the Rapid Burster evidently works on the principle of filling up an energy reservoir to some critical level before it springs a leak and gives a flash of X-rays. The neon flasher receives energy by being electrically charged; the Burster draws on the gravitational potential energy of accreted matter in the Lamb model discussed earlier (p. 114). The near-constancy of the peak burst intensity also suggests that there may be a self-regulating device at work – a device which may work through the effect of radiation pressure and heating on the motion of the plasma falling through the leak in the reservoir. In the massive black hole model it becomes difficult to account for these features of the Rapid Burster.

The Rapid Burster operates in two modes. Mode I is 'seen' in March and September every year, while mode II follows in April and October. Mode I represents large X-ray bursts with energy in the range 10^{32}–10^{33} J together with small bursts in the energy range 10^{31}–10^{32} J. Mode II represents predominantly bursts of energy 10^{32} J. The bursts are further divided into two types: type I and type II. In a type I burst the spectrum* of X-rays softens as the intensity drops; that is, the proportion of lower energy photons in the total radiation progressively increases. In a type II

---

* In a *spectrum* of X-rays the intensity of radiation of X-ray photons of differing energies is specified.

burst the shape of the spectrum stays fixed. The two modes of X-rays in the Rapid Burster are of type II, but the Burster occasionally emits type I bursts of energy exceeding 10^{32} J. These bursts are not so frequent; they occur once every few hours.

While considering theoretical models for the type I bursts from the Rapid Burster, S. M. Chitre and K. M. V. Apparao of the Tata Institute of Fundamental Research in Bombay arrived at the conclusion that these bursts in X-rays should also be accompanied by bursts of *infrared* radiation. In their model for X-ray bursts Chitre and Apparao followed the general idea of leakage of the plasma reservoir from the magnetosphere onto the surface of the neutron star. However, as the plasma descends through the magnetic field near the polar regions of the star, interesting side-effects take place.

Instead of simply falling on the star surface and releasing X-ray bursts, the protons and electrons of the plasma behave differently. As the protons fall they encounter strong resistance from the growing magnetic field. To be able to descend further the protons must get rid of their magnetic moment: a task which they are able to achieve by transferring their transverse momentum (i.e. momentum perpendicular to the direction of the magnetic field) to the electrons which are also falling. This momentum-transfer can take place through collisions, with the result that the electrons acquire extra energy. All this happens at a distance of some fifty times the neutron star radius.

The feature which becomes important from the view point of infrared emission is the large momentum of the electrons perpendicular to the magnetic field lines. This situation, leading to an anisotropic distribution of electrons, promotes the operation of a *cyclotron maser*. Cyclotron radiation is like the synchrotron radiation described in Appendix E, only it concerns radiation by electrons moving at speeds slow compared to the speed of light. In a maser we have an inversion of population. In a normal population of electrons moving at random the low energy electrons predominate over the high energy ones. In a maser, which represents a coherent process, this distribution is inverted – leading to more high energy electrons. Such a coherent process seems quite plausible and should result in the production of infrared bursts if matter were to descend on to the poles in an intermittent fashion.

Prompted by this suggestion Chitre and Apparao persuaded the infrared group at the Physical Research Laboratory, Ahmadabad, to search for infrared bursts from Liller I. P. V. Kulkarni and N. M. Ashok from this group accordingly used their liquid-nitrogen cooled photometer with a set of infrared filters at the focus of the 1 metre telescope of the Indian Institute of Astrophysics at Kavalur in South India. On the night of

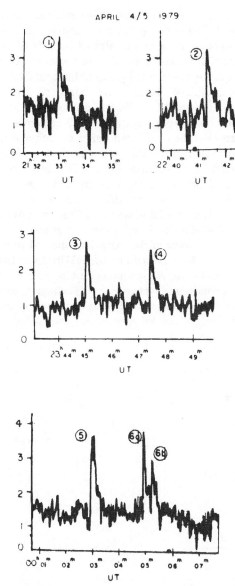

Fig. 6.6. The six infrared bursts seen from Liller I. The ordinate is in units of 10^{-16} watts cm^{-2} micron^{-1} while the time axis records Universal Time. (By courtesy of P. V. Kulkarni, N. M. Ashok, K. M. V. Apparao, and S. M. Chitre; reprinted from *Nature* **280**, 819 (1979).)

4, 5 April 1979 they were rewarded with six infrared bursts from the direction of Liller I. The burst profiles are shown in Fig. 6.6.

The estimated energy in a typical infrared burst is of the order of 10^{31} J – about the right order expected from the model of Chitre and Apparao. Subsequently the infrared group at Imperial College, London under the direction of A. W. Jones, in collaboration with their Spanish colleagues C. S. Margo and M. P. Munoz, also saw two similar infrared bursts during their observations of Liller I on 5 September 1979.

This is the first example of infrared bursts to date and it would be interesting to see whether there really is a connection between the X-ray bursts and the infrared bursts from Liller I. For this, it is essential to monitor the source simultaneously in the X-rays and in the infrared. If the two are simultaneously turned on and turned off, it will be strong evidence in support of the model advanced by Chitre and Apparao and of the leaking plasma models for X-ray bursters in general.

This concludes our survey of the more dramatic and powerful X-ray sources in the Galaxy. As indicated first by UHURU and later by more advanced X-ray satellites, the X-ray sources exist in profusion beyond our Galaxy also. In the following chapter we discuss extragalactic examples of violent phenomena and amongst these X-ray emission will feature again.

7 Active Galaxies and Quasars

'I am convinced, my dear Watson, that a black hole is responsible for this violent act.'

'A black hole!' I exclaimed incredulously.

'Yes, a supermassive one. How often have I said to you that when you have eliminated the impossible, whatever remains, *however improbable*, must be the truth?'

(With apologies to Sherlock Holmes)

A large ground-based telescope can probe the Universe out to distances of the order of ten thousand million light years. Within this range, it is estimated that there lie hundreds of millions of galaxies like our own. Our Galaxy, a system of some 10^{11} stars contained in a disc shaped distribution of radius $\sim 50\,000$ light years, is typical of disc shaped galaxies, which constitute over 50 per cent of the total number of galaxies. In a typical disc galaxy we may expect, besides the stars, some 10 per cent of the matter in the form of gas and a still smaller percentage in the form of dust. The stars may be going round the centre of the galaxy in much the same way that the planets move round the Sun. In our Galaxy, it is estimated that a star may typically take around 200 million years to complete one orbit.

A closer examination of the discs of such galaxies reveals further structure. Figure 7.1 shows examples of a few varieties. The spiral structure so evident in these objects has been responsible for the name '*spiral galaxies*' often given to such systems.

There is, however, another class of galaxies known, again because of their shapes, as '*ellipticals*'. They are generally ellipsoidal in shape and consist of relatively old stars moving in more complicated orbits than the stars in disc galaxies. These galaxies further differ from the disc types in that they are almost devoid of gas and are also without any overall rotation so common to disc galaxies. Also, compared to the disc galaxies, the ellipticals have a larger mass to light ratio. This is because they consist mainly of old stars which are well past their most bright phases. Thus to produce the same brightness more stars are necessary in an elliptical galaxy than in a disc galaxy. The ellipticals are estimated to constitute about 10 per cent of áll galaxies.

Although considerable work has been done in the last two decades to study the morphology of galaxies, the astronomers are still far from an acceptable theory of how galaxies form and how they evolve. This is not to say that theories don't exist! They do; but they have not reached the same level of sophistication that theories of star formation and stellar evolution have attained.

In this book we are not concerned with the 'silent majority' of normal galaxies of stars. Our concern here is with the so-called *active* galaxies which show some kind of violent activity either within themselves or in their vicinity. These exceptional members stand out amongst the rest and have naturally attracted special attention from the astronomers and the astrophysicists. We begin our discussion of extragalactic violent phenomena with M 87, a giant elliptical galaxy in the *Virgo* cluster of galaxies.

Messier 87

The *Messier* catalogue of bright objects in the sky was compiled by Charles Messier (1730–1817). The Crab Nebula discussed in Chapter 5 is the first object of this catalogue. Just as the Crab exhibits many remarkable properties amongst the galactic objects, M 87 is one of the most striking objects lying outside our Galaxy.

In Fig. 7.2 and Fig. 1.8 we see some photographs of this galaxy. In a long-exposure photograph (Fig. 7.2) the object looks spherical with a diameter of about 22 500 light years, small compared to the ∼ 100 000 light year diameter of the disc of our Galaxy. In the short-exposure photograph (Fig. 1.8), however, a peculiar feature of the galaxy stands out, namely a *jet* coming out of its nucleus (the *nucleus* of a galaxy means its dense central region). The jet is about 4000 light years long and its existence has been known from photographs taken six decades ago.

The photograph 1.8 does not convey the gigantic size of M 87, for it does not show the halo of light which surrounds the spherical region, a halo of diameter ∼ 650 000 light years! Moreover, the radio astronomers find M 87 to be a strong emitter of radio waves. We will return to the radio properties of M 87 when we consider radio sources in general. M 87 is also the first X-ray galaxy to be detected. In 1970 Friedman and his group at the Naval Research Laboratory detected the X-rays from this galaxy. The X-rays come from the jet as well as from a halo surrounding the galaxy. The X-ray halo is even bigger than the light halo: its diameter is over 800 000 light years! A schematic diagram of M 87 with its various radiating parts is shown in Fig. 7.3.

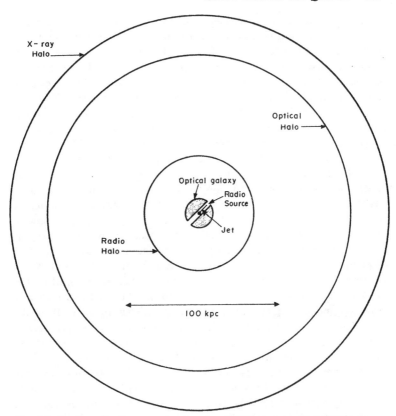

Fig. 7.3. A schematic diagram showing the different parts of M 87. The large optical and X-ray halos indicate that the galaxy is different from normal galaxies. 1 kpc $\simeq$ 3262 light years.

What could be the cause of X-ray emission from such a gigantic region? Of the various processes described in Appendix E the *inverse Compton process* seemed at first the best suited to account for the X-ray emission from the halo of M 87. In the inverse Compton process a light photon collides with a fast moving electron. In the collision the photon gets most of the energy from the electron. The result is that the electron slows down while the photon increases its frequency. Under suitable conditions the light photon may increase its frequency so much that it becomes an X-ray photon.

So in the case of M 87 we need two ingredients to generate the X-radiation: a source of fast moving electrons and a source of light photons. The electrons may originate in the source itself through the explosion which gave rise to the jet, for example. But where is the source of photons?

In the mid-1960s Fred Hoyle suggested that the source of photons could be none other than the cosmic microwave background radiation whose existence was first detected in 1965 by A. A. Penzias and R. Wilson. We will return to a discussion of this radiation in the next chapter. It is sufficient to say here that the intensity of this radiation in the halo of M 87 exceeds by far the light which could come from the stars of that galaxy. Thus the microwave photons are better candidates for generating X-rays through the inverse Compton process than the light photons in M 87. The gigantic halo of X-rays around M 87 therefore seemed to arise from the interaction of very fast electrons with the cosmic background radiation.

While the inverse Compton process is a viable mechanism for production of X-rays in general, it now seems unlikely that this process could be responsible for the X-rays in M 87 or in other X-ray sources found in clusters of galaxies. The recent observations using X-ray spectroscopy have revealed the existence of emission lines associated with iron, oxygen, and other elements which clearly point to the mechanism of thermal bremmstrahlung discussed in Appendix E as being the source of X-ray emission. In this process we have hot intra-cluster gas trapped by the strong gravitational pull of the galaxies in the cluster. The presence of iron, oxygen, etc. indicates that this gas is enriched by heavy elements. The X-rays are emitted when electrons and ions collide. The presence of typical ions can be identified from specific frequencies of the X-ray photon. Thus X-ray emission with photons of 1.1 keV energy has been identified with the presence of iron, in much the same way that an optical astronomer can identify emission lines due to various elements in the spectrum of visible light. It would not be possible to understand this line emission in the inverse Compton model.

Improvements in X-ray observations brought about by the Einstein Observatory have added further confirmation to this picture. It is now possible to have detailed contour maps showing how the X-ray brightness of a galaxy like M 87 varies as we go away from the central region. If the mechanism for X-ray emission were related to the density of hot gas trapped near the galaxy, as the thermal bremmstrahlung process requires us to assume, we should find a correlation between the intensity of X-ray emission and the central galaxy density. Such a correlation has been found.

M 87 is a dominating member of the Virgo cluster and the X-ray emission from it is characteristic of radiation from a cluster of galaxies. For this reason M 87 was listed in Table 6.1 amongst X-ray sources of class G: the clusters of galaxies. It is believed that X-ray emission from clusters is due to the inverse Compton scattering of microwave radiation.

Just as the halo of M 87 has proved to be so interesting, its central nucleus has also provided considerable food for thought for theoreticians.

Figure 7.2 does not really reveal the details of the nucleus. Even Fig. 1.8 which reveals the existence of the jet cannot tell us much about the central nucleus. Indeed until a few years ago the astronomers could not say anything more specific than that the nuclear region is a seat of some violent activity.

It was with considerable interest, therefore, that in 1978 astronomers greeted the results of two independent investigations of the central region of M 87. One group of investigators, consisting of P. J. Young, J. A. Westphal, J. Kristian, and P. Wilson of the Hale Observatories and F. P. Landauer of the Jet Propulsion Laboratory, examined the brightness distribution across the inner parts of the galaxy. The second group included W. L. W. Sargent and P. J. Young from the Hale Observatories, A. Boksenberg and K. Shortridge from University College, London, C. R. Lynds from the Kitt Peak National Observatory, and F. D. A. Hartwick from the University of Victoria. This group made a spectroscopic study of the galaxy.

Both groups arrived at more or less the same conclusion, that their observations could be best explained by the hypothesis that the nucleus of M 87 contains a *supermassive black hole*! ·

This remarkable conclusion was not new to the theoreticians. In 1963 Fred Hoyle and William A. Fowler had suggested gravitational collapse as a means of forming supermassive objects in galactic nuclei. In 1966 Hoyle and this author suggested that massive objects (with mass $\sim 10^8 \, M_\odot$) may exist in nuclei of elliptical galaxies and that these masses would influence the distribution of stars in the ellipticals and their luminosity profiles. In 1969 D. Lynden-Bell first suggested that a supermassive black hole may exist in a galactic nucleus or in a quasar and generate energy.

However, to return to observations! Of the two investigations, the first group used two very accurate detector systems, known as SIT and CCD* on the 60 inch and 200 inch telescopes at Mt Palomar. The angular resolution of the data was of 1″ (that is, the measurements could distinguish points separated by as small an angle as 1″) and the sensitivity of light collecting capacity was of the order of 1 per cent. Such high accuracy had not been achieved before and this is why for the first time a detailed study of the brightness distribution of the galaxy could be carried out as far inwards as 0.17″ from the central nucleus. Figure 7.4 shows how the luminosity of M 87 increases as we go inwards from an angular distance of 80″.

If we assume that the visual light of the galaxy comes from the stars

---

* SIT stands for the Silicon Intensifier Target television camera tube used in the system. CCD is the Charge-Coupled Device based on semiconductor technology.

within it, then the distribution of brightness across its surface can be calculated from the distribution of stars in it. Models of how stars distribute themselves in a spherical system governed by one another's gravity had been constructed in 1966 by I. King, and these models had provided a reasonable fit to the earlier data on luminosity profiles of elliptical galaxies including M 87. However, the present data extended closer to the nuclear region than before and it is here that the discrepancy from the King model was noticed.

In Fig. 7.4 the dotted line shows how the luminosity should rise towards the centre in the King model and the continuous line describes how in fact it is seen to rise in M 87. The agreement which is good in the outer parts (shown by a thick line) where the two lines coincide, breaks down within a distance of 1″ from the nucleus. The observed profile rises above the predicted profile indicating an *excess* of stars in the nuclear region. How

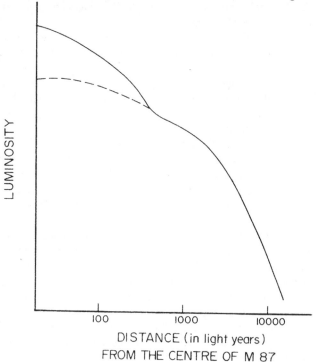

Fig. 7.4. The luminosity profile of M 87 is shown by the continuous line, while the dotted line shows the expected profile from standard star distribution models. The excess observed in the central region is attributed to a supermassive black hole. Luminosity drops by a factor 100 from the centre to the outer part of the above figure.

could this excess of stars be drawn towards the central regions? After examining several likely alternatives the investigators put their money on the hypothesis that there exists a supermassive black hole in the centre of the galaxy, pulling the stars inwards. They estimated the mass of the black hole to be $5 \times 10^9 \, M_\odot$.

The second group of investigators made a spectroscopic study of M 87 using the instrument called IPCS* developed at University College, London. The spectral data can be subjected to a mathematical analysis called the Fourier analysis. This analysis gives us the information about *velocity dispersion* of matter emitting the light.

Figure 7.5 illustrates the principle behind the analysis. In (a) we see a 'sharp' spectral line, of the type emitted by a fixed source of light. The 'line' is in fact a sharply peaked curve showing that most of the energy is coming in the form of light of a particular frequency v. In (b) we see a somewhat broader curve, indicating a spread in the range of frequencies over which the light from the source is distributed. The peak is still at v but the spread around v shows that because of the Doppler effect (see Appendix C) the light from the different components of the source has undergone different shifts in frequency. Thus if some parts are moving towards us their light is peaked at a slightly higher frequency than v. Similarly, if some parts are moving away from us their light peak occurs at a lower frequency than v. A superposition of differently peaked profiles produces the broader curve of Fig. 7.5(b). The larger the overall spread in the velocities the larger the line broadening.

The second group of investigators of M 87 found this broadening effect to increase as they turned their attention from the outer to the inner

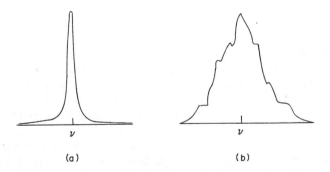

v v

(a)   (b)

Fig. 7.5. In (a) we see a narrow line profile. In (b) the same line is widened due to velocity dispersion discussed in the text.

* Short for **Image Photon Counting System**.

regions. A larger broadening implies, as we just saw, a larger spread in source velocities – measured by the so-called *velocity dispersion*. This velocity dispersion in M 87 rose from $\sim 230\ \mathrm{km\,s^{-1}}$ at the angular distance of 72″ from the centre to $\sim 350\ \mathrm{km\,s^{-1}}$ at the angular distance of 1.5″. The sharp rise in velocity dispersion again indicates a strong gravitating source in the central region of M 87. The mass of the source estimated this way is again $\sim 5 \times 10^9\ M_\odot$.

These M 87 investigations, both coming at the same time, have given a strong boost to the idea of a supermassive black hole. Although other interpretations of the data cannot be ruled out at this stage, the black hole interpretation was considered the simplest – provided one can say how the black hole formed in the first place! Further, it must be pointed out that the present investigations go as far as 0.17″ from the centre, which at the estimated distance of M 87 corresponds to a distance of ~ 40 light years. Thus all we can conclude on the basis of these results is that there exists a mass of $5 \times 10^9\ M_\odot$ within a spherical region of radius ~ 40 light years. For this mass to be a black hole, its radius must be as small as 1.5×10^{-3} light years! Hence we have still a long way to go before we can assert that the object in the nucleus of M 87 is a black hole and not an ordinary collapsed object (with radius larger than 1.5×10^{-3} light years) which happens to be faint.

As a postcript it is worth adding that an alternative model *not* requiring a massive black hole in the centre of M 87 has already been proposed by J. C. Wheeler and M. J. Duncan. These authors have shown that both the light profile and the rising velocity dispersion can be better explained by assuming that the stars in the central region are moving in an anisotropic fashion. Clearly the last word has not been said on the black hole picture!

Radio Astronomy

Let us now turn our attention to another branch of astronomy which also gives ample evidence for violence on a large scale. Like X-ray astronomy, the preliminary investigations of radio astronomy were motivated by commerce rather than by science. In 1930 the Bell Telephone Laboratories at Holmdel, New Jersey commissioned Karl Jansky to investigate the possible sources of atmospheric interference which might hamper the working of a ship to shore communications system that the laboratory was planning to operate. The working frequency of the equipment was 2×10^7 cycles per second. The radio astronomer nowadays refers to frequency in units of *hertz* (to commemorate H. R. Hertz who first demonstrated during 1885–9 the existence of electromagnetic waves by a laboratory experiment). Thus hertz (shortened to Hz) stands for one cycle per second and a more suitable unit for radio astronomers is the *megahertz* (= 10^6 Hz), denoted briefly by MHz.

Working at the frequency of 20 MHz, Jansky built a revolving antenna 30 m wide and 4 m high. Although with his equipment Jansky was able to locate various local atmospheric disturbances, he also made a discovery of astronomical significance. He found radio waves coming from the direction of the Galactic Centre.

This discovery was made in 1933. Any further steps in radio astronomy had to wait till the end of the Second World War. In 1944 G. Reber, who had built a large fixed antenna in his back garden, was able to construct the first radio map of the Galaxy, showing that there was a concentration of radio waves in the direction of the Sagittarius constellation. There were also smaller peaks in the directions of Cygnus and Cassiopeia.

It was J. H. Oort, the then Director of the Leiden Observatory in Holland, who was the first amongst professional astronomers to see the significance of Jansky's discovery. Instead of a background noise, Oort felt the need for a specific radio frequency which could be used to probe astronomical objects, just as spectral lines are useful to the optical astronomer. His young student H. C. van de Hulst succeeded in 1944 in locating such a frequency, at 1.42×10^9 Hz, corresponding to the wavelength of about 21 cm, arising from transitions in hydrogen atoms. Later Oort and C. A. Muller instituted a search for radio waves at this wavelength in the Galaxy and in 1951 they found such waves. A week or two earlier H. L. Ewen and E. M. Purcell had also succeeded in detecting a 21 cm line in gas clouds in the Galaxy.

Our interest in this chapter is in extragalactic astronomy. So let us, in this brief historical narrative, move on directly to the discovery of the first radio source outside the Galaxy. In 1946 J. S. Hey, S. J. Parsons, and J. W. Phillips made the discovery of the source now known as Cygnus A. In 1951 F. G. Smith was able to locate the position of Cygnus A accurately enough to enable the optical astronomers to identify a visible astronomical source with it. (We have encountered this procedure of *optical identification* in connection with the X-ray sources in Chapter 6.) The identification was made by W. Baade at the Mount Wilson and Palomar Observatories, who found that the source coincided with an object photographed by Baade and R. Minkowski in 1953. The object is shown in Fig. 1.9.

This object at first sight looks like a couple of galaxies close to each other and consequently the early interpretations of radio sources were based on the concept of *colliding galaxies*. It was felt that a radio source like Cygnus A is a rare but powerful event which requires an extraordinary explanation. Collision of two galaxies seemed to fill the bill. However, as we will see in the following section, something even more dramatic and powerful is required to account for the behaviour of radio sources.

In the 1950s England and Australia took the lead in setting up radio telescopes and in making observations of extragalactic interest. The systematic attempts to catalogue extragalactic radio sources like Cygnus A were made by Sir Martin Ryle and his colleagues at Cambridge and by B. Y. Mills and his colleagues in Australia.

The United States entered the field in a big way in the 1960s with the setting up of the National Radio Astronomy Observatory at Green Bank, West Virginia, and the 1000 foot dish at Arecibo, Puerto Rico. As technology improved, different types of radio telescopes began to be built with improved sensitivity and improved ability to locate a source and to make detailed structural studies of sources. In very long baseline interferometry (VLBI) radio astronomers can link telescopes thousands of kilometres apart to make very detailed observations of radio sources. These observations have probed the sources down to the scale of a light year and have led to some very remarkable results. We will discuss one aspect of the VLBI observations when we come to quasars.

Extragalactic Radio Sources

What were the reasons for abandoning the colliding galaxies hypothesis for radio sources like Cygnus A? The reasons were twofold. First, in the late 1950s, G. R. Burbidge gave an elegant argument estimating the energy store in a strong radio source like Cygnus A. Burbidge's calculations made use of one observed property of the radio waves coming from Cygnus A: *the waves were polarized.*

In any electromagnetic wave the electric and magnetic disturbances are at right angles to each other and at right angles to the direction in which the wave propagates (see Appendix A for details). However, it is not necessary that the waves present in a given radiation all possess the same electric and magnetic directions. In general these directions may be distributed randomly and such radiation is called unpolarized. If, by some special arrangement, these directions are kept the same then we have a *plane polarized* radiation. The radio waves from many cosmic sources are polarized, thus indicating some special arrangement in the way they are produced.

In Appendix E we have discussed a number of ways in which electromagnetic radiation may be generated. Only one of these processes is capable of generating polarized waves. This is the process of *synchrotron* radiation.

In the synchrotron process, radiation is generated when a fast moving electric charge is made to change its direction by a magnetic field. The radiation so produced is polarized with the direction of the magnetic field in the wave made parallel to the original magnetic field. The frequency of radiation and its intensity depend on two factors: the strength of the magnetic field and the energy of the fast particles.

From the available data on the frequency spectrum and the intensity of radiation from a typical radio source, Burbidge was able to estimate the *minimum* energy that must necessarily be present in the particles and in the magnetic field in order to sustain the observed radiation by the synchrotron process. Typical total energies in these two modes are comparable and work out to a staggering figure of $\sim 10^{55}$ J – about 10^{40} times the energy released in the explosion of a megaton H-bomb.

How much energy can a pair of colliding galaxies produce? The collision hypothesis depended on the conversion of the gravitational potential energy of the colliding pair of galaxies into energy of radio waves. That is, in the process of collision, the gravitational energy is expected to generate fast particles which would then radiate by the synchrotron process. For two galaxies of masses M_1 and M_2 separated by a characteristic distance R the gravitational energy is of the order

$$E = \frac{GM_1 M_2}{R}.$$

For $M_1 = M_2 = 10^{11} M_\odot$ and $R \simeq 150\,000$ light years we get $E \simeq 10^{52}$ J, a thousand times *less* than the requisite energy! Thus dramatic though a collision of galaxies is expected to be it is not powerful enough to sustain radio sources like Cygnus A.

In the 1960s the optical and radio observations of Cygnus A and other radio sources had begun to reveal a different type of picture altogether. A typical radio source is like that described schematically in Fig. 7.6. It has a

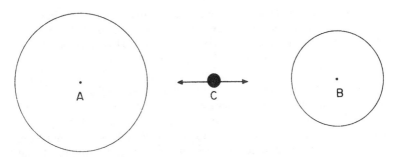

Fig. 7.6. A typical double radio source with two radio emitting regions A and B. The radio emission is associated with emission of plasma from the central region C, often associated with a galaxy.

double structure, being made of two radio emitting blobs separated by a distance of the order of 500 000 light years. Compare this with the diameter of our Galaxy which is of the order of 100 000 light years and note that radio sources much more extended than the above 'typical' size also exist, and you get some idea of the immensity of these systems! In Cygnus A the two blobs may be as far apart as a million light years.

Optical data show that usually there is an active galaxy somewhere in the middle, along the line joining the two blobs, as shown in Fig. 7.6. The galaxy associated with Cygnus A, shown in Fig. 1.9, is similarly located. In some cases, analysis of the particle motion in these galaxies shows large outward velocities indicating some kind of explosion. There is *no* evidence for a system of colliding galaxies.

Any modern theory of radio sources must take account of the double structure, the central explosion, and the large energy reservoir needed for keeping the sources radiating. To date, no satisfactory energy machine has emerged. We describe below a chain of ideas considered to be the 'best buy' amongst the existing theories.

Let us recall the reason for failure of the collision hypothesis: the quantity GM_1M_2/R was not large enough to energize the reservoir of charged particles to the requisite values. This quantity could be increased if either M_1, M_2 were made larger or R made smaller. Now M_1, M_2 are already as large as galactic masses and it would be difficult to increase them further. On the other hand R can be made smaller if we could imagine more compact systems than galaxies. We also recall from earlier chapters that for a compact object we can improve the performance of gravity by increasing the ratio M/R so much that

$$\frac{2GM}{c^2R} \sim 1.$$

Thus to have efficient and large production of energy from a gravitational source we need a massive compact object close to the black hole state. For these reasons theoreticians nowadays feel that a *supermassive black hole* is likely to provide the answer to the energy production of radio sources.

This conclusion as such is not new. As far back as 1963, F. Hoyle and W. A. Fowler suggested the gravitational collapse of a supermassive object as the source of energy of strong radio sources. In the late 1960s Philip Morrison and A. Cavaliere suggested the idea of a *spinar*: a supermassive rotating object (like a neutron star but $\sim 10^8$ times more massive!) whose gravitational energy is converted to rotational and magnetic energy and thence to radiation.

That gravity *can* supply the requisite energy is known; the problem for the theoreticians has been to cook up a credible scenario whereby the gravitational energy is converted to electromagnetic radiation in an efficient manner!

More recently, in 1974, Martin Rees and Roger Blandford proposed another version of this process which involves accretion on to a rotating supermassive black hole. The accretion process is similar to that discussed for X-ray sources. A disc of accreted material is formed round the black hole. In Box 7.1 we show how the black hole mass is estimated from the observed data on radiation. A black hole of mass $\sim 10^8 \, M_\odot$ can account

---

Box 7.1 Supermassive black hole as an energy machine

We saw in Box 4.1 that the Penrose process enables one to extract ~ 30 per cent of the total energy residing in a black hole. The various theories employing supermassive black holes use this property of the black hole implicitly. Thus the total energy residing in a black hole of $10^8 \, M_\odot$ is

$$10^8 \, M_\odot c^2 \sim 2 \times 10^{55} \text{ J}.$$

Even if 10 per cent of this energy is available for conversion to optical or radio emission, this would be sufficient to explain the energy reservoirs of quasars and strong radio galaxies.

Next, if we use the accretion disc scenario from Box 6.2, we get for $\dot{M}_0$ = accretion rate $\simeq 1 \, M_\odot$ per year and for $M = 10^8 \, M_\odot$

$$E \simeq 10^{-1.75} \text{ keV},$$

which corresponds to a wavelength of about $700 \, \text{Å}$ ($= 7 \times 10^{-8}$ m). This lies in the ultraviolet range. A smaller accretion rate would result in the main radiation being in the optical part of the spectrum. These figures are suitable for quasars which emit strongly in the ultraviolet and the optical.

Finally, the Eddington luminosity (cf. Box 6.2) for $M = 10^8 \, M_\odot$ gives a value of $\sim 10^{38}$ w – again in the right region for quasars.

---

for the observed radiation of extended sources and of quasars, to be discussed later. The efficiency of energy conversion from gravity to electromagnetic radiation is claimed to be as high as 20 per cent. (Compare this with the efficiency of energy conversion in the hydrogen burning stage of the main sequence star which is only 0.7 per cent.)

The problem remains, how do we get an exact alignment and the double structure of Fig. 7.6? Rees and Blandford have proposed a 'twin exhaust' model illustrated in Fig. 7.7. In this model plasma is squirted out from the flat disc in the directions of least resistance. For a rotating system these directions lie along the axis of rotation. The flow pattern of the two jets in Fig. 7.7 is like that of jet engines and known in aerodynamics as '*de Laval nozzle*'. The plasma is squirted out in two opposite directions and it proceeds until it encounters resistance from the interstellar medium which

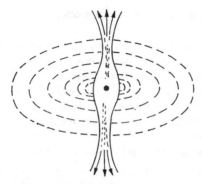

Fig. 7.7. The twin exhaust model. Plasma is squirted out along the two nozzles aligned with the axis of rotation.

limits the distance it can go and hence the size of the double radio source. The radio emitting blobs are explained as regions where the magnetic field is present leading to radiation by the synchrotron process.

It will be possible to make further progress with radio source models only after we have detailed maps of these sources such as are becoming available by the VLBI techniques and by the VLA (Very Large Array) recently completed at New Mexico in U.S.A. Only then will it be possible to settle such questions as whether the fast plasma in the blobs acquires all its speed in the central explosion or whether it is further energized *in situ*. Also whether one needs just a collapsed object like a spinar, or a black hole, is not so far clear from the theoretical discussions.

Quasars

The first quasars were discovered early in 1963 through a collaboration between radio astronomers and optical astronomers. By now more than 1300 of these remarkable objects are known, and some of their observed characteristics are as follows.

First, they are star-like in appearance showing sharp point-like images with emission lines in their spectra. Indeed the first two quasars, 3C 48 and 3C 273, were mistaken for stars until their spectra were examined more closely. This examination showed, what is now known to be a common property of all quasars, that the quasars have large red-shifts. The red-shift of 3C 273 was 0.158 and that of 3C 48 was 0.367. To date, the largest red-shift record is held by the quasar OQ 172, with a red-shift of 3.53. Some quasars also show absorption lines in their spectra, which are also red-shifted. In some of these cases there are differences in the emission line red-shift and absorption line red-shift of a quasar (in some quasars there are multiple absorption line red-shifts).

Indeed the interpretation of quasar red-shift has been a subject of controversy which we will not discuss here. The 'establishment view' is that the quasars are very distant objects and that their red-shifts arise *solely* from the expansion of the universe in a manner to be described in Chapter 8. We will return to a 'non-conformist view' of red-shifts in Chapter 9.

Although the quasars were originally discovered through radio astronomy and many of the new ones still are found from their radio properties, the property of radio emission is believed to be present in only a few per cent of all quasars. That is, the majority of quasars are radio quiet.*

The progress of X-ray astronomy has revealed another facet of quasars. Most quasars are strong X-ray emitters. Although a hint of this property was given by the early X-ray observations of 3C-273, the more recent observations of the Einstein Observatory (HEAO-B) have shown about 100 quasars as strong X-ray sources. Unlike the radio emission, X-ray emission is believed to be a common property of quasars on the basis of these early observations. This discovery has posed a new problem to which we will return at the end of this chapter.

The most difficult problem connected with quasar models is how to explain their energy production. If the quasars are at cosmological distances (that is, if their red-shifts arise from the expansion of the Universe), then the above problem amounts to explaining how a luminosity as high as 10^{39} watts – about a hundred times that of our Galaxy – originates in a volume of a few cubic light years. A view held by many astrophysicists is that quasars are linked to galactic nuclei in an evolutionary process. We have seen, for example, how the nucleus of M 87 is considerably brighter than its outer parts. There are other types of galaxies, known as Seyfert galaxies (see, for example, Fig. 1.7), in which the contrast between a bright nucleus and fainter surroundings is even more marked than it is in M 87. The quasars may be one step further on this sequence so that if they are located at large distances we only see the bright central nucleus and nothing of the faint periphery (if it exists at all).

This is the establishment view and it is not difficult to extrapolate from this and argue that, as in M 87, quasars may also contain supermassive black holes serving as energy machines. Indeed, since the mid-1970s, the idea that a quasar is powered by a supermassive black hole has gained considerable popularity.

Once it was decided that a supermassive black hole *is* responsible for quasar luminosity, the theoretical bandwagon could move in two

---

* Originally, quasars were called 'quasi-stellar radio sources'. Since many quasars are radio quiet, the name is now changed to 'quasi-stellar objects'. The short version 'quasar' is also often used, although in the early days it was not considered respectable!

directions. In one direction lay the problem of explaining how a supermassive black hole could form in the first place, while the other direction faced the problem of finding a process which can convert the black hole's gravitational energy into radiation energy. Some progress has been made on both these fronts.

Various scenarios have been constructed leading to the formation of a supermassive black hole. For example, a gigantic cloud may straightaway collapse to a black hole. Or it may condense into a star cluster which evolves through star collisions, gas depletion, and disruptions into a supermassive star which then becomes a black hole. Another method makes use of the formation of massive ($\sim 100\ M_\odot$) stars which become supernovae and leave behind neutron stars or stellar mass black holes which may subsequently form the supermassive black hole. The aim of these and various other processes is to argue that beyond a certain stage gravity dominates over all disruptive processes and pulls in matter to form the black hole.

The energy production process also relies on the pulling power of gravity. The 'best buy' idea relies on accretion round a rotating black hole, as in the case of extended radio sources. R. D. Blandford and R. L. Znajek have suggested a dynamo process in which the magnetic field also plays a part. The magnetic lines of force are dragged along by the rotating black hole. The dragging, however, is not perfect: slippage occurs. That is, the magnetic lines slip across the rotating disc which has formed round the black hole, and in this process they generate an electric field. This electric field is in the direction of the axis of rotation and it forces electrically charged particles to move along the axis in opposite directions. This process therefore supplies a scenario in which charged particles are fired from the central region of a quasar in opposite directions. The radio emission from quasars (which also show double structures like the extended radio sources described earlier) therefore arises from the synchrotron emission by these charged particles in ambient magnetic fields.

The X-ray emission from quasars could arise from the accretion disc. Recently a quasar OX 169 showed a drop in its X-ray emission rate by more than a factor of 3 in just $2\frac{1}{2}$ hours. This implies that the emission region cannot be more than a few light hours across, i.e. no more than about 10 times the size of a supermassive black hole of mass $10^8\ M_\odot$. The concept of Eddington luminosity also leads to a black hole mass of this order (see Box 7.1).

We now close this chapter by outlining two problems associated with quasars. One relates to their internal structure and the other to their X-radiation.

Faster than Light Motion?

As mentioned before, the VLBI techniques in radio astronomy revealed detailed structure on the scale of light years. One remarkable result of such measurements was the discovery that the central region of a double radio source itself has several components on a considerably smaller scale. In particular, in many cases it was found that there are two blobs in the central region which themselves are moving away from one another. The VLBI techniques could not only measure the angular separation between these blobs, which was of the order of milliseconds of arc, but they could also measure changes in it from year to year.

Suppose the blobs are located from us at a distance D and suppose their angular separation as seen from here is α. If the blobs were lying perpendicular to the line of sight, their physical separation from each other is simply $D \times \alpha$. If they do not happen to lie in this special direction, the separation must be even larger than $D \times \alpha$. Let us assume the separation to be $D \times \alpha$. The VLBI measurements measure the rate of change of α. Therefore we can estimate the rate of separation of the two components to be

$$v = D \times \text{rate of change of } \alpha.$$

The observations of a few quasars in the 1970s showed the startling result that v computed in this way exceeded the speed of light c: in some cases v was as high as 20 c! Was this result in direct contradiction to Einstein's special relativity theory?

A direct resolution of this problem is of course to argue that our estimate of quasar distance D may be wrong. The distance has been estimated using Hubble's law, to be described in Chapter 8. This law presupposes that the quasar red-shift is due to the expansion of the universe and an estimate of its distance according to this law is given by multiplying the red-shift by a fixed number. If, however, this estimate is wrong and the quasar is considerably nearer, then the above calculation of v is also wrong. The value of v, the separation velocity, is then much less than that calculated above and certainly does not exceed c.

To many astronomers firmly dedicated to the belief that Hubble's law applies to quasars, this resolution of the difficulty is not very appealing. So, many contrived scenarios have been suggested to show that the observed motions are not real but illusory.

One scenario makes use of the '*Christmas tree*' effect. In an illuminated Christmas tree the electric lamps are arranged to go on and off in a sequential fashion generating the impression of motion. A similar illusion of motion is generated by neon signs used for advertizing. The observed separation of quasar components might similarly be an effect produced by the activation of radio components along an axis in a sequential fashion!

A second scenario, based on an effect first pointed out by Martin Rees in 1966, makes use of the fortuitous circumstance that the two radio blobs in the quasar may be lying nearly along the line of sight. In this case, if they are moving away from the central region with speeds comparable to c then their apparent separation speed can appear to exceed c. This effect is based on, and is consistent with, Einstein's special relativity.

Another scenario, described by S. M. Chitre and the author, explains the effect with the help of gravitational bending. In Fig. 7.8 we see two components A and B observed by the distant observer O. If along the line of sight there exists a massive object, its gravity may bend the radio waves in the manner described in Chapter 3. The objects are therefore not 'seen' at A and B but at A' and B'. Thus it is likely that even though A and B do not move apart with a relative speed exceeding c, their images A' and B' may well do so. Under special circumstances this velocity magnification can be very high.

This theory is capable of being tested. A direct test may be the detection of a massive object along the line of sight to the quasar. The object may be as massive as a galaxy, but if it is considerably fainter than a typical galaxy

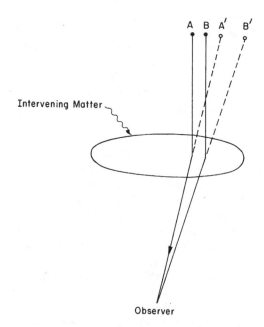

Fig. 7.8. By gravitational bending the images of A and B are formed at A' and B'.

it may be difficult to detect. The space telescope to be launched in 1985 is capable of detecting objects 50 times fainter than those which can be seen by ground-based telescopes. It may well be able to search for such an intervening object.

The bending of light by an intervening object is not altogether improbable! In fact a strong case in favour of gravitational imaging of a quasar by an intervening cluster of galaxies was recently made in a different context, and it is worth discussing this briefly.

In 1979 D. Walsh, R. F. Carswell, and R. J. Weymann discovered two quasars, now labelled 0957 + 561 A and B, with the identical red-shift of 1.41 and very similar spectral features. The similarity of the two quasars together with their nearness in the sky (they were separated by the very small angle of 5.7 arc second) led to the speculation that these two objects were the images of *one* quasar located somewhere between them. It was argued that a galaxy along the line of sight to the quasar may bend the light rays from them in such a way as to produce more than one image. A search for such a galaxy finally led P. Young, J. E. Gunn, K. Kristian, J. B. Oke, and J. A. Westphal to a probable candidate: a galaxy of red-shift 0.39 located in a cluster of galaxies with an angular separation of about 1 arc second from B.

Young and his colleagues have argued that this galaxy, together with the surrounding cluster, acts as a gigantic gravitational lens which not only produces the observed images A and B but also reproduces the broad radio maps made of these quasars.

Although the faster than light motion has proved embarrassing to quasar watchers, it has held out hope for the proponents of the supermassive black hole theories of quasars. For the separation of the blobs in the central region is found to be well aligned in the same direction as the outer blobs of the double radio source. Could this be evidence that plasma is continuously being squirted out along the axis of rotation of the black hole?

The X-ray Background

One of the major discoveries of the Einstein Observatory has been the detection of profuse quantities of X-rays from quasars. The early observations suggest that the X-ray luminosity of the quasar is correlated to its optical luminosity: the brighter the quasar in visual light, the brighter it is in X-rays. A bright quasar like 3C 273 can emit X-rays at as high a rate as 10^{39} w.

These observations initially created a difficult situation which can be illustrated by the following example.

Imagine a football ground being lit by flood lights at night. The overall light generated on the ground depends on the number of flood lights used and their intrinsic power. However, suppose that in spite of many powerful lights the ground is still in relative darkness! How can this happen? This is precisely what was suspected in the case of X-ray quasars. If we estimate the total X-ray background generated by quasars only, the result comes out higher than what is actually observed! To this one must add the contributions from ordinary galaxies and from clusters of galaxies which also happen to generate substantial quantities of X-rays. Thus the discrepancy is made worse! To resolve the discrepancy we have to suppose either that quasars do not exist in as much profusion as previously believed or that by and large they are less powerful than those first detected by the Einstein Observatory. It now seems that the number of quasars was over-estimated and that the background problem looks like being resolved. The fact that the problem existed for a while is a testimony to the important input X-ray astronomy is capable of adding to our overall view of the Universe.

8 The Big Bang

Echoing the dissatisfaction of many intellectuals about the concept of the 'Big Bang', the Hungarian mathematician Paul Erdos says: 'God committed two acts of folly. First, He created the Universe in a Big Bang. Second, He was negligent enough to leave behind evidence for this act, in the form of the microwave background.'

The subject of *cosmology* deals with the large scale structure of the Universe. It aims very high in that it seeks to answer questions regarding the origin, the evolution, and the future behaviour of the Universe as a whole. Although in the past philosophers have conjectured, poets imagined, religious leaders claimed to know the answers to these questions, today the subject of cosmology comes under the discipline of science. And today the majority of cosmologists would assert that the Universe was created in a gigantic explosion, the *big bang*, and that it has been evolving since its birth.

Since our purpose throughout this book has been to describe instances of violent activity in the Universe, we can do no better than conclude our discussion with the most violent activity ever imagined. In this chapter we will stick to the conventional picture – the picture based on Einstein's general relativity and the laws of physics known today. In the next and the last chapter we will relax some of the bonds of conventionality in order to seek answers to some questions which have remained unanswered within the strictly conventional framework.

The Expanding Universe

Two series of investigations, one observational and the other theoretical, both occurring during 1915–30, went a long way towards establishing cosmology as a branch of science. The observational stimulus came from the work of E. Hubble, J. H. Oort, M. Humason, and H. Shapley and the theoretical stimulus from A. Einstein, W. de Sitter, and A. Friedmann.

Right until the end of the first decade of this century the astronomers were firm in their belief that our Galaxy, encompassed by the Milky Way band, constituted the *whole* of the observable Universe. Indeed our Galaxy was regarded as the unique 'island' in the vast 'sea' of space. There were a few doubters like R. A. Proctor who advanced the contrary view that the observable Universe did not stop at our Galaxy but extended well beyond it. But these sceptics were as usual outnumbered by the conservative majority.

However, improved telescopes with better auxiliary equipment began to show that the majority view was wrong. Careful observations showed how some of the diffuse nebulae seen across the Milky Way were in fact entire galaxies situated well beyond our own. An instance of this change of attitude may be cited here. For example, it was argued that there was a significant absence of faint nebulae in the plane of the Milky Way. If the faint nebulae were outside the Galaxy, their distribution in the sky should be the same in all directions. Hence astronomers used to conclude that these nebulae cannot be outside our Galaxy. Gradually it became clear that the apparent absence of the nebulae in the direction of the plane of the Milky Way was due to the interstellar absorption, and not due to the actual absence of these objects.

It was Hubble who with his colleague Humason at the Mt Wilson Observatory (now part of the Hale Observatories) in California made the remarkable discovery referred to in Chapter 1. The absorption lines in the spectra of some of the nearby galaxies showed *red-shift*.* We have already come across the notion of red-shift in two different forms: the Doppler red-shift and the gravitational red-shift. In the 1920s general relativity was still to be established as a respectable theory; at any rate it had not become as widely known and understood as it is today. Hence it was natural for these observers to interpret the red-shifts of galaxies as arising from the Doppler effect. This effect is described in Appendix C where we note that if the red-shift z is small compared to unity, we may assign a velocity of recession

$$v = cz$$

to the source.

Hubble also found that the red-shift seemed to increase with the faintness of the galaxy. If we assume two galaxies G_1 and G_2 to be equally luminous and if G_1 *appears* to us to be fainter than G_2, then we conclude that G_1 must be located further away from us compared to G_2. Indeed the astronomer, in this way can relate the faintess of a galaxy to its distance. Hubble, therefore found that the red-shift z of a galaxy increases with its distance D from us according to a simple formula:

$$v = cz = HD.$$

* To remind the reader: if λ is the observed wavelength of a spectral line and λ_0 the expected wavelength, then the red-shift z is given by $(\lambda - \lambda_0)/\lambda_0$. Thus for a positive value of z, the observed wavelength must *exceed* the expected wavelength.

Here H is a constant known as *Hubble's constant*. It measures the increase in the recession velocity v per unit distance. Hubble's original estimate in 1929 was that the velocity increases by about 160 km per second for every million light years of distance. (Later we will find another way to state the value of the Hubble constant.) Subsequent work has shown that Hubble had grossly underestimated the distances of nearby galaxies, with the result that the above value of Hubble's constant is too large by a factor of between 5 and 10. Thus the present value of Hubble's constant may be as low as between 16–32 km per second for every million light years. The reason for this uncertainty in stating the exact value of H is that even today astronomers do not agree amongst themselves on what distances to ascribe to nearby galaxies. To fix ideas we will take H at around 22.5 km per second per million light years.

A *first* assessment of Hubble's law seemed to suggest that there may have been an explosion in the neighbourhood of our Galaxy as a result of which all galaxies are flying apart from us. Although there is no *prima facie* objection to such an interpretation, it has the drawback of making our Galaxy somewhat special in the Universe. Like physicists, astronomers do not like to postulate special conditions in order to account for their observations. At any rate since the days of Copernicus any hypothesis ascribing a privileged status to the local observer has been anathema to the theoreticians. Fortunately by 1929 cosmological theory had made sufficient progress to account for Hubble's observations *without* having to grant any special status to the Galaxy. This is therefore the appropriate stage to go back to 1915 and pick up the story from the theoretician's angle.

In 1915 Albert Einstein proposed his theory of gravity, the so-called general theory of relativity. In Chapters 3 and 4 we discussed the applications of this theory in the neighbourhood of a massive object. We have seen how the geometry of space-time is modified by the gravitational influence of a massive distribution of matter in the form of a star like the Sun, or in a black hole, say. In 1917 Einstein made a more daring application of his theory. He applied it to the Universe as a whole. Since the Universe is a conglomeration of massive objects like galaxies, the *overall* geometry of the Universe is expected to be different from that of Euclid's.

As mentioned earlier, in 1917 Einstein had very little to go by so far as actual cosmological observations are concerned. He therefore opted for the *simplest* model of the Universe that could be thought of in those days. The Einstein model, in technical terms, is described as being *homogeneous*, *isotropic*, and *static*. These technical aspects can be understood in the following way if we imagine galaxies as the typical vantage points for viewing the Universe. Then *homogeneity* implies that, seen from any galaxy, the Universe presents the same large scale view. *Isotropy* means that, if we look at the Universe from any one of the galaxies, all directions present the same large scale view. Finally, a *static* Universe does not show any systematic large scale motion of its constituents, the galaxies. Thus it looks the same at all instants of time if viewed from any galaxy. Notice how the first property takes care of the Copernican doctrine: no galaxy has a privileged position in the Universe. The second property removes any special status for any direction, while the last property makes all moments of time of equal significance.

Einstein's trouble with this highly symmetrical Universe was that he could not obtain it as a solution of his own 1915 equations! So he proceeded to *modify* his equations by introducing a *new* force of nature. This is the λ-force which tells us that there is a cosmic force of *repulsion*, proportional to their separation, acting between any two masses in the Universe. The constant λ measures the strength of this force. The constant is so small that on the scale of the Solar System, or of typical stars, the repulsive force is negligible compared to gravity. Its effect really shows up on the scale of the Universe, on distances of the order of 10^9–10^{10} light years!

Einstein had hoped that his cosmological model would show explicitly how the distribution of matter determines the characteristic features of the non-Euclidean geometry of the Universe. This expectation came out to be true. The Einstein Universe has a finite volume which is related to the density of matter. Einstein also expected that his model would turn out to be unique, in the sense that no other simple model would follow from his modified equations. However, in 1917, in the same year that Einstein proposed his model, W. de Sitter proposed another model – from the very same equations used by Einstein. De Sitter's model describes an empty Universe which is homogeneous and isotropic but not static! In fact, by 1924 A. Friedmann had demonstrated that homogeneous and isotropic but non-empty cosmological models could also be obtained from the original Einstein equations of 1915. These models, like the de Sitter model, are not static. They show the Universe to be *expanding*.

The Universal expansion can be understood in the following way. Suppose, as in Fig. 8.1, we have a triangle $G_1 G_2 G_3$ of three galaxies, the observers on which have the ability to measure the distances $G_1 G_2$, $G_2 G_3$, and $G_3 G_1$ at various instants of time. In Fig. 8.1 we see how this triangle would compare with itself over different instants of time. As the time progresses the triangle gets bigger and bigger in size. This effect would apply to a triangle made of *any* three typical galaxies. In other words, the space between galaxies is expanding.

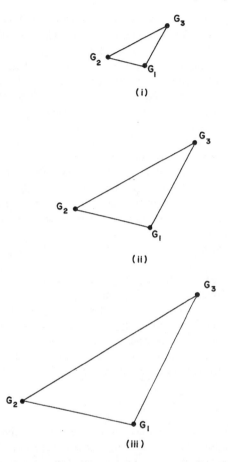

Fig. 8.1. The triangle $G_1\,G_2\,G_3$ of galaxies is shown at three epochs (i), (ii), and (iii) in chronological order. The triangle 'expands', being twice and thrice the original size of (i) in the later epochs (ii) and (iii) respectively.

How can we as observers confined to *one* galaxy measure this effect? Certainly for us to conduct such measurements of triangles of different galaxies extending over time spans of thousands of millions of years is plainly impossible. The Friedmann models nevertheless hold out another prospect of testing this effect. If the triangles in Fig. 8.1 are expanding, then seen from any vertex the other two vertices should appear to recede, because the two sides of the triangle emanating from that vertex are growing in size.

Now we go back to the observations of Hubble and his colleagues and recall that they had found just such an effect! The red-shift from a typical galaxy does indicate that it is receding from us, if we use the simple Doppler effect interpretation. Theoreticians and observers in the early 1930s could put these facts together to build up models of the Universe which were the mathematical solutions of Einstein's equations and which at the same time were consistent with Hubble's observations.

However, once we decide to work within the framework of general relativity, it is desirable that we use the interpretation of that theory to describe the phenomenon of the red-shift. This is done by first recognizing that there exists a scale factor S which changes with time and which tells us how fast the Universe is expanding. In our Fig. 8.1, the scale factor measures the relative sizes of the sides of the triangles. Thus if we take the earliest triangle to correspond to $S = 1$ then the next triangle has $S = 2$ and the one after that $S = 3$. This means that the sides of the second (third) triangle are twice (thrice) the size of the corresponding sides of the first triangle.

Now suppose light from galaxy G_1 left at the first epoch and was received by G_2 at the second epoch in Fig. 8.1. Then G_2 will find the wavelengths of the various spectral lines in this light to have *doubled* since they left G_1. Thus he will conclude that G_1 has a red-shift $z = 1$. In general if the light left the source when the scale factor had a value S_1 and is received by the observer when the scale factor has the value S_2 ($> S_1$), then the source has a red-shift z given by the simple formula

$$1 + z = \frac{S_2}{S_1}.$$

This result follows from the considerations of how light propagates in the non-Euclidean space-time of the expanding Universe. If we identify t_2 with the present epoch, then it is often convenient to refer to the past epoch t_1 as the epoch of red-shift z. The mathematical analysis of the above relation also tells us how Hubble's law follows from it. The Hubble constant H receives the following simple interpretation. At any epoch when H is measured, its value gives us the rate at which S is increasing at that time *divided by* the value of S at that time.

To summarize: the Friedmann models retain all but one of the symmetries of the original Einstein Universe. These models are not static, but they are homogeneous and isotropic at any cosmic epoch. In fact, the homogeneity and isotropy of a Friedmann model enable us to identify a universal time. Two observers sitting on different galaxies can match their cosmic clocks by comparing the physical properties of the Universe in their respective neighbourhoods. For example, they can identify the instants when the density of matter of one neighbourhood agrees with that in the other. That the Universe should be homogeneous and isotropic at any cosmic epoch is known as the '*cosmological principle*'.

The Big Bang

The fact that the Hubble constant is positive means that S is *at present* increasing with time. What was its behaviour like in the past? What will it do in the future? Questions like these cannot be answered entirely on the basis of present observations. Mathematical models such as the Friedmann models described above do provide us with a framework within which to assess the answers to such questions.

Figure 8.2 shows how the expansion of the Universe proceeds with time in different types of Friedmann models. The continuous curves marked I, II, and III represent three such typical models. All curves have been arranged to touch one another at point P which describes the *present* epoch. At P we have arranged the three curves I – III to have the same value of S and the same value of Hubble's constant H. The present epoch is denoted by t_p.

In these models, as we look towards the future of t_p, curves I and II continue to show increasing scale factor implying that the Universe continues to expand *forever*. Curve III, on the other hand, indicates that the scale factor increases up to a certain epoch t_c after which it decreases to zero. Accordingly the Universe will continue to expand up to the epoch t_c after which it contracts. This contraction is very similar to the contraction of a massive body undergoing a gravitational collapse (see Chapter 4).

The curve III describes a *closed Universe* whereas the curves I and II describe *open Universes*. If we go back to our discussion of Chapter 3 and recall that space can have different types of curvature, we now see that these three models provide us with examples of those curvature types. Curve I has a space of zero curvature, curve II has a space of negative curvature, while curve III has a space of positive curvature. The volume of space in models of type I and II is *infinite* while that in the type III Universe is *finite*. To visualize a type III Universe it helps to think of a lower dimensional space. The surface of a sphere has two dimensions. It has a finite area but no boundary. A flat man wandering on this surface may make a round of the surface and return to his starting point. The type III Universe has similarly a finite volume of space but no boundary. A ray of light from any point in space may make a round of space and return to its starting point. The Einstein Universe described earlier had the spatial geometry of type III.

Although the three models differ in these respects, they share one common property which is apparent when we look to the *past* of t_p. All

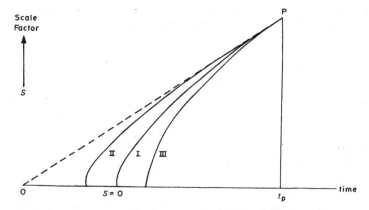

Fig. 8.2. The scale factor S in three typical Friedmann models I, II, and III. The dotted line shows the case in which S increases at a fixed rate.

three curves hit the line $S = 0$ at some time in the past. That is, all Friedmann models predict that the Universe had an epoch in the past when all its contents were compressed in a space of *zero volume*. How long ago was this remarkable epoch?

The dotted tangent to the three curves at P gives an upper limit to this time interval. The tangent meets the time axis at the point 0. The time interval from the epoch of 0 to the present epoch is simply $1/H$. As seen from Fig. 8.2, all three curves hit the time axis at epochs after 0. This is because these curves tell us that the rate of expansion has been slowing down with time, that it was larger in the past than it is now. The dotted line, on the other hånd, has been obtained on the assumption that the rate of expansion was the same in the past as it is now, which is why this line hits the time axis at the most remote epoch of 0.

Whichever Friedmann model we choose to accept as a working theory, it is clear that it predicts that there was an $S = 0$ epoch in the past. It is customary to take this epoch as the beginning of the Universe and to count time from that epoch. Accordingly we note that the 'age' of the Universe – i.e. the time elapsed since the $S = 0$ epoch to the present epoch – is the least in the type III model and the largest in the type II model and in any case less than $1/H$.

Now we recall that the constant H was introduced as the ratio of velocity to distance. So the dimensions of $1/H$ are the dimensions of time. Our value of $H = 22.5$ km per second per million light years leads us to the value of $1/H \simeq 13$ thousand million years. This is the *upper limit* to the age of the Universe.

It is clear that we can fix the actual age of the Universe if we could decide which is the correct Friedmann model, if any! Although considerable observational efforts have gone into settling this question, an unambiguous answer has so far eluded the cosmologist. We will not go into the various observational tests proposed so far,* but will concentrate our attention on the epoch $S = 0$.

The epoch of $S = 0$ is a singular epoch, of the type we encountered towards the end of Chapter 4. At this epoch the mathematical description of space-time geometry breaks down, a feature now known to be common to most physically realistic solutions of Einstein's equations. The existence of such singularities suggests some inadequacy with general relativity. It may well be that a future theory will be free from such a 'fault'.

Some cosmologists accept the singularity as something to be expected since the creation of the Universe is a highly special event. If nothing existed prior to this epoch, then it is clear that the creation phenomenon represents a wholesale violation of the law of conservation of matter and energy. Such a breakdown of physics can only be justified by assuming the existence of a singularity.

To me this attitude indicates an abdication of responsibility by the physicist. To argue that the laws of science have a limited domain of applicability is alien to the spirit of science. It is very likely (I think it is a certainty) that our physics has not yet acquired the maturity to deal with epochs when S was small, and the existence of singularity simply reflects this inadequacy. This does not mean that physics will never be able to explain the basic problem in cosmology, the problem of describing where the matter and energy we see around us came from in the first place. All I am advocating is caution, that we should not be confident that our simple extrapolations of the limited physical knowledge at present definitely answer the ultimate question in cosmology.

Physicists have nevertheless pushed their investigations as close to the singular epoch as they can dare to do! The purpose of such investigations has been to try to see if any signature of the early times has survived to this day. For, as we shall see in the following discussion, the closer we approach the epoch $S = 0$, the more turbulent we find the state of the Universe. The naming of this epoch as the 'big bang' epoch is intended to convey this state of affairs in the early Universe.

Primordial Nucleosynthesis

At the epoch of big bang the Hubble constant is infinite and so is the average density of matter in the Universe. As the Universe expands both

---

* See for a detailed discussion of this subject *The Structure of the Universe* by this author (Oxford, 1977).

the Hubble constant and the matter density steadily decrease. The matter density ρ falls off in later stages as

$$\rho \propto \frac{1}{S^3}.$$

If we recall our definition of the red-shift we can express the above result in the following form. Let ρ_p be the present matter density and ρ the matter density in the past epoch of red-shift z. Then

$$\rho = \rho_p (1 + z)^3.$$

Thus if we observe a galaxy of red-shift $z = 1$, we are seeing it at an epoch when the matter density in the Universe was *eight* times its present value. The further we look into the past the larger is the value of z and the larger is the value of ρ. This dependence of ρ on z is illustrated in Fig. 8.3.

In Fig. 8.3 log ρ is plotted against log $(1 + z)$ giving us a straight line of slope 3. In the same figure we have plotted the logarithm of radiation density u against log $(1 + z)$. This line is steeper, having the slope 4, indicating that the radiation density at the epoch of red-shift z is related to

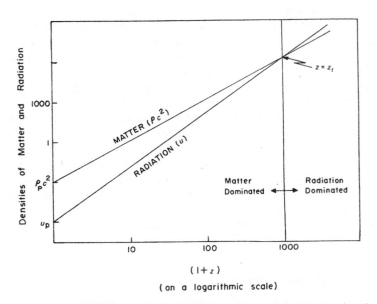

Fig. 8.3. The logarithm of density ρ of matter and u of radiation plotted against the logarithm of $(1 + z)$. The two curves intersect at a transition value of red-shift z_t $\simeq 1000$. The Universe is radiation dominated for $z > z_t$ and matter dominated at $z < z_t$. (The factor of 1000 rise in density is indicated on the density scale.)

that at the present epoch by the relation

$$u = u_p (1 + z)^4.$$

Thus at the epoch of $z = 1$, the radiation density was *sixteen* times the present radiation density.

The matter density may consist of stars in galaxies, gas and dust in galaxies, and clusters of galaxies, quasars, etc. There may also be invisible matter in the form of black holes. More recently the exciting possibility has emerged that neutrinos may have a non-zero mass, and thus massive neutrinos created soon after the big bang may by now have shed most of their energy of motion and come to rest. Such neutrinos, if they exist, will also contribute to ρ_0. What about radiation density?

We know that visible light is just one aspect of electromagnetic radiation. In the last two chapters we saw that sources of radio waves, infrared, and X-rays also exist in abundance in the Universe. Thus we expect to have radiation present in all these forms. However, as we shall shortly see, the dominant form of electromagnetic radiation observed presently in the Universe is not in any of the above forms, but is in microwaves. According to the present estimates the microwave background contributes most to u_p.

So far as the relative importance of matter and radiation in the expansion of the Universe are concerned, we may compare the two by using Einstein's relation $E = Mc^2$. Thus we have to compare u, the radiation density, with ρc^2 in deciding which is more effective in controlling the rate of expansion of the Universe.

What are the present estimates of u_p and ρ_p? u_p is estimated at $\sim 10^{-14}$ J per cubic metre. The estimate of ρ_p is more tricky. If we confine ourselves to matter actually seen in the form of galaxies, quasars, etc., the estimate of ρ_p is $\sim 10^{-28}$ kg/m^3. There may, however, be hidden or non-luminous matter such as black holes, massive neutrinos, and other material in the intergalactic space as yet not fully explored: and hence the above estimate of ρ_p could be an under-estimate. Nevertheless, even this under-estimate, used in Fig. 8.3, is such that $\rho_p c^2$ exceeds u_p by a factor ~ 1000.

Looking at Fig. 8.3 we note that the two lines intersect at $z \simeq 1000$. This is the epoch at which the radiation density and matter density (expressed in energy units) are equal. That such an epoch must have existed in the past is clear from the fact that radiation density rises faster than matter density as z increases. We denote by z_t the transition epoch at which $u = \rho c^2$. At redshifts exceeding z_t, the radiation density dominates over matter density. In fact, as we approach the big bang epoch the radiation completely dominates over matter.

It was George Gamow who first appreciated the likely importance of radiation in the early stages of the big bang Universe. The radiation is expected to be in the black body form (see Appendix E) and has a temperature T given by

$$u = aT^4$$

where a is the radiation constant. It therefore follows that as we go towards the big bang epoch through successive epochs of increasing redshifts, the radiation temperature increases as

$$T = T_p(1 + z).$$

Here T_p is the present day radiation temperature.

When we use this fact of radiation domination in the early stages of the Universe, Einstein's equations give us a very simple relationship between the temperature T and the time t elapsed since the big bang. The relation is simply:

$$T = \alpha . \frac{10^{10}}{\sqrt{t}} K.$$

Here T is measured on the absolute scale and t in seconds. The constant α is of order unity and it depends on the actual state of matter and radiation present in the Universe. If we take $\alpha = 1$, we see that the temperature of the Universe was 10^{10} K one second after the big bang.

It was Gamow's idea that such a hot oven in the early Universe would be able to 'cook' different nuclei, starting from a relatively simple primordial soup of electrons, positrons, neutrinos, antineutrinos, neutrons, and protons besides the photons of electromagnetic radiation. During the period 1946–50, Gamow and his co-workers R. Alpher, H. Bethe, and R. Hermann carried out extensive calculations to demonstrate the feasibility of this idea.

In retrospect we can now say that Gamow was only partially right in this idea. Nucleosynthesis in the primordial era *is* possible but it cannot deliver all the elements observed in the Universe. The primordial process outlined below can give us light nuclei like deuterium and helium only. The heavier nuclei like carbon, oxygen, etc. have to be obtained in stellar nucleosynthesis, the details of which we considered in Chapter 5. The difficulty lies in the fact that there is a gap of unstable light nuclei like lithium which cannot be surmounted in the nuclei-building process in the early Universe. The synthesis therefore terminates at this stage, essentially after helium is formed.

Gamow's calculations were redone in the 1960s and 1970s with fresh inputs from nuclear and particle physics. Special mention may be made of the work of C. Hayashi from Japan, Ya. Zeldovich and

I. Novikov from the U.S.S.R., F. Hoyle and R. J. Tayler from the U.K., and R. V. Wagoner, W. A. Fowler, and P. J. E. Peebles from the U.S.A. The scenario of primordial nucleosynthesis has been summarized in Table 8.1.

Table 8.1 Primordial Nucleosynthesis

Age of the Universe	Temperature (K)	State and composition of matter
10^{-2} second	10^{11}	n, p, e$^-$, e$^+$, $\nu, \bar{\nu}$ in thermal equilibrium with n and p in equal numbers
10^{-1} second	3×10^{10} K	Same particles as above but n : p ratio $\simeq$ 3 : 5
1 second	10^{10}	$\nu, \bar{\nu}$ decouple from the rest while e$^-$, e$^+$ begin to annihilate; n : p $\simeq$ 1:3
13.8 second	3×10^9	D and ^{4}He begin to form. e$^-$, é$^+$ disappear. Free n, p also present.
35 minutes	3×10^8	The abundance of D and ^{4}He fixed relative to protons and electrons. ^{4}He/H ratio $\sim$ 22–28 % by mass is now frozen.
7×10^5 years	3×10^3	Chemical binding strong enough to form neutral atoms. The Universe is now transparent to radiation and is also becoming matter dominated.

Key to symbols: n (neutron), p (proton), e$^-$ (electron), e$^+$ (positron), ν (neutrino), $\bar{\nu}$ (antineutrino), D (deuterium), H (hydrogen), He (helium).
Adapted from *The Physics–Astronomy Frontier* by F. Hoyle and J. V. Narlikar (W. H. Freeman, 1980).

The table lists the sequence of events in a chronological order starting from the epoch when the Universe was one hundredth of a second old. From our formula relating temperature to time we see that the Universe at this stage has a temperature of around 10^{11} K. The Universe is now supposed to contain predominantly light particles, there being about one neutron or proton for 10^9 photons, electrons, and neutrinos! The neutrinos do not interact with other matter except through weak interaction (see Appendix D) and in an ordinary terrestrial environment neutrinos and antineutrinos can travel through matter essentially undisturbed. However, in the primordial era the high density and temperature make even the neutrinos interact. Thus the following interactions take place:

$$\bar{v} + p \rightleftharpoons e^+ + n,$$
$$v + n \rightleftharpoons e^- + p.$$

The reactions above proceed both ways leading to frequent absorption and emission of neutrinos (v) and antineutrinos ($\bar{v}$). The whole distribution of particles is, however, in a thermal equilibrium with the high temperature of 10^{11} K.

The temperature of this cosmic soup drops with time and by the time the Universe is 1/10 second old it is around 3×10^{10} K. At this stage a property which governs the overall distribution of various particles in thermal equilibrium begins to have an important effect on the sequence of events. According to this property the relative numbers of various particles are weighted by a rule which takes note of the particle energies. Thus the more energetic particles are less abundant than the less energetic ones. Again, if we use the Einstein relation $E = Mc^2$, we see that the more massive particles will be less favoured than the less massive ones. The effect on the distribution of these energy (or mass) differences become significant when the overall temperature T drops to a low enough value at which kT becomes comparable to these energy differences. Here k is a thermodynamic constant known as the Boltzmann constant.

At the temperature of $T = 3 \times 10^{10}$ K the quantity kT equals an energy of about 2.5 MeV, which is comparable to the energy difference between a proton and a neutron, both at rest. Being slightly more massive the neutron is now less abundant than the proton. At the age of 1/100th of a second the neutrons and the protons were nearly equal in number; now there are about 3 neutrons to every 5 protons.

As the temperature drops to 10^{10} K at the age of 1 second, another change begins. The neutrinos can no longer be confined to the rest of matter through the above interaction. They decouple themselves and begin to cool down as a separate entity. The neutron–proton ratio now drops to about 1:3. The electron–positron pairs begin to annihilate each other so that their total number is lowered while the number of photons (arising from the annihilation) increases.

The temperature drops to about 3×10^9 K when the Universe is nearly 14 seconds old. At this stage the attractive nuclear force between the proton and the neutron begins to be strong enough to hold them together. Thus are formed the nuclei of deuterium (1 neutron + 1 proton) and helium (2 neutrons + 2 protons). We will return to these processes after completing our survey of Table 8.1.

The nuclei-building processes terminate when the temperature is further lowered to around 3×10^8 K at the age of 35 minutes. That means there will be no more fusion of neutrons and protons into these nuclei. When this happens the proportion of helium (^{4}He) to free protons lies in the range 0.22 to 0.28 by mass. The number of electrons has also come down to the level of equality with the number of protons.

From here on considerable time has to elapse before the next significant development occurs. At the age of, say, 7×10^5 years, the temperature falls to 3000 K. The chemical binding between atomic nuclei and the electrons is now strong enough to form neutral atoms. Thus hydrogen and helium atoms form by this epoch.

Here we notice one coincidence. With the disappearance of free electrons into atoms the main agency which was responsible for scattering the radiation is removed. Because the radiation is no longer scattered it can travel long distances. In other words, the Universe becomes transparent to radiation. The epoch when this happens turns out to have a red-shift of ~ 1000 – roughly the transition epoch separating the radiation dominated phase from the matter dominated phase! For, if we recall how the radiation temperature depends on the red-shift of the epoch, we find that a temperature of 3000 K at red-shift of 1000 corresponds to a temperature close to 3 K at the present epoch. Just such a temperature has in fact been observed! We will return to this remarkable observation in the following section.

We end this section by referring back to the formation of helium and deuterium. There is some evidence to suggest that the abundance of helium in the Universe is as high as 25 per cent by mass. The evidence is *not* incontrovertible and the 'universality' of the abundance is claimed on the basis of observations of some stars in our Galaxy and of some neighbouring galaxies. The issue of whether the figure of ~ 25 per cent is universal (and not just typical of a few types of astronomical objects) is crucial to the theory of primordial nucleosynthesis. For, the only other way of producing helium is through stellar nucleosynthesis and this process can produce only ~ 8 per cent of helium by mass. If stars were required to produce more helium, there has to be more hydrogen burning and hence more light produced by the main sequence stars. This in turn would make galaxies including our own considerably brighter than they are now known to be. Hence a high universal helium abundance would make the primordial nucleosynthesis much more credible than it otherwise would be. There is, however, a loophole in this argument which is discussed on p. 157.

The deuterium abundance in cosmic material is no more than $\sim 2 \times 10^{-4}$ by mass. Yet even such a small quantity of deuterium seems difficult to produce in ordinary stars. The primordial nucleosynthesis process, on the other hand, can account for this abundance provided the present density in the form of nucleons is no more than a few times 10^{-28} kg/m^3. Again the crucial issue is whether the observed instances of deuterium abundance can be considered representative of the universal abundance.

Although the issue is not yet settled, it is potentially a test which could tell us whether a big bang origin of the Universe is a necessary hypothesis in cosmology. If the deuterium observations do suggest a big bang origin, they will also imply that the Universe must be open, not closed. For the Friedmann models tell us that if the Universe is to be closed (i.e. of type III) its density must exceed

$$\rho_c = \frac{3H^2}{8\pi G}.$$

For our value of H and the measured value of the gravitational constant G, this density ρ_c (known also as the *closure density*) must be as high as 10^{-26} kg/m^3. Thus the deuterium test, if confirmed, would rule out closed universes, unless they contain substantial non-nucleonic matter.

The Microwave Background

We have seen in the previous section that considerations of the early Universe lead us naturally to the conclusion that we should expect to see a low temperature radiation background as the relic of the hot big bang. Such a prediction was in fact made by Gamow and his co-workers in the late 1940s. In those days the theory of primordial nucleosynthesis had not attained the level of sophistication it has today and this prediction was not precise enough to be able to forecast the exact value of the temperature. Gamow's quoted value of 5 K was more an inspired guess than based on a detailed theory. A black body radiation with temperature of this order is predominantly in the microwave form.

Curiously enough Gamow's prediction was not taken seriously; at any rate nobody bothered to look for such a radiation in the 1950s although microwave detection technology had developed to a level at which this measurement could have been made. In the early 1960s there was a renewed interest in primordial nucleosynthesis and other aspects of the early Universe. R. H. Dicke and P. J. E. Peebles at Princeton arrived at conclusions similar to Gamow's, and realizing the significance of measuring the temperature of the background radiation they persuaded their experimentalist colleagues P. G. Roll and D. T. Wilkinson to set up an antenna to measure the intensity of background radiation at the wavelength of 3.2 cm.

The first measurement of this radiation, however, did not come from the Princeton group. It came early in 1965 and quite by accident, from

A. A. Penzias and R. W. Wilson at the Bell Telephone Laboratories, Holmdel, New Jersey. Penzias and Wilson were attempting to estimate the 'noise' from the Galaxy at the microwave wavelength of 7.35 cm with the help of an antenna using a 20 foot horn reflector of low noise built for communication via the Echo satellite. Once again, as in the examples cited on pp. 97 and 126, equipment designed for commercial use yielded a result of paramount scientific significance. For Penzias and Wilson found a radiation background which was *isotropic* in nature and which corresponded to a black body temperature of ~ 3.5 K. Not being aware of the cosmological implications of the microwave background, Penzias and Wilson puzzled over their result for several months. However, when the news of their discovery reached Princeton, Peebles could put two and two together. What Penzias and Wilson had found appeared to be the relic radiation from the big bang.

The Princeton group shortly followed with their own measurement at 3.2 cm. The two points did fall, within their experimental errors, on a black body curve of temperature ~ 3 K. Subsequently many other measurements were made and these are shown in Fig. 8.4. At the top we see the ground-based observations which are mostly on the long wavelength side of the expected black body peak. For a few years many technological hurdles had to be surmounted before measurements at the short wavelength side could be made from above the atmosphere. The figure below shows the result of such measurements made by D. P. Woody and P. L. Richards. Although there are large experimental errors (shown by the shaded regions) the overall pattern of distribution of radiation does resemble a black body curve. The 'best fit' black body curve has a temperature of 2.96 K, although statistically the fit cannot be considered very good. Does this imply that there is a genuine discrepancy between the theoretical prediction and actual observations? Should one look for small secondary effects to explain the observed discrepancies? Or should one just wait until more refined measurements are made before taking the discrepancy seriously? Theoreticians have reacted differently to these questions, reflecting their own personal prejudices.

Although the majority of astronomers today take the view that the microwave background measurements provide a direct 'proof' of the hot big bang, there have been a few dissenters who would like to explore other alternatives. Here we outline a few of them briefly.

One approach is based on the observed 'coincidence' that the energy density of the microwave background ($\sim 10^{-14}$ J/m^3) is comparable with (if slightly higher than) the energy density of starlight, the galactic magnetic field, and of cosmic rays. But, of these four phenomena, the last three are of relatively local character and of relatively recent origin. Only the

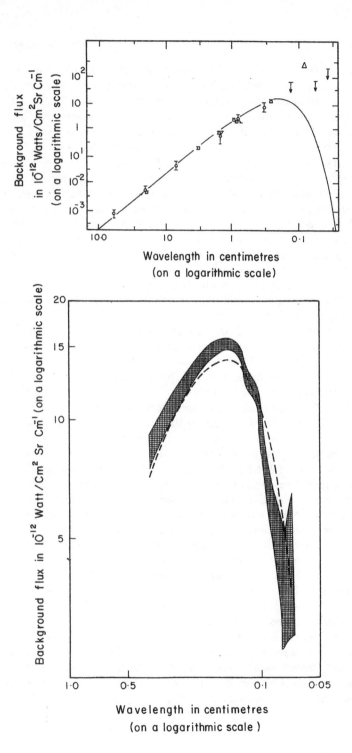

microwave background stands as the odd one out if it is interpreted as being of big bang origin. Hence some theoreticians have tried to think of astrophysical processes of relatively recent origin to account for the microwave background. One such process is the process of thermalization of other radiation by intergalactic dust grains. In a typical thermalization process the dust grains absorb other radiation (e.g. starlight, infrared and ultraviolet radiation, etc.) and reradiate it largely in the microwaves. The observed isotropy of the microwave radiation, and its near black body nature, puts severe restraints on what type of intergalactic grains can do this job. The original suggestion came in 1975, jointly from N. C. Wickramasinghe and M. J. Edmunds of University College, Cardiff and S. M. Chitre, S. Ramadurai, and the author from the Tata Institute of Fundamental Research (T. I. F. R.), that whisker shaped grains of graphite might be suitable agents for this process. More recently in 1980 N. C. Rana from T. I. F. R. has argued that pyrolytic graphite may be more suitable than natural graphite. It still remains to be seen how this idea stands up to the various theoretical and observational constraints.

In 1977 M. J. Rees from the Institute of Astronomy at Cambridge advocated the use of dust grains in a different context to explain the microwave background. In contrast to the 'hot' big bang discussed so far, Rees considered the 'cold' big bang! That is, he took the view that the Universe need not have passed through an early hot era and that the microwave background observed today is not the relic of a very hot era just after the creation of the Universe. The radiation could have originated much later. In his theory Rees argued that even before the galaxies formed from diffuse clouds by condensation there were supermassive objects of mass $\sim 10^6 M_\odot$. Such objects may have formed when the Universe was about a million years old and they would have evolved rapidly, as massive stars do, passing through their most active phase in ten million years. They would synthesize nuclei which would be ejected in supernovae, as discussed in Chapter 5. This leads to a substantial amount of dust in the medium which can act as a thermalizer of light. In the picture proposed by Rees gas molecules and the ionized matter also play a role. Note also that because these supermassive stars operated *before* the galaxies formed we can get the observed helium from these stars without requiring the galaxies to be brighter in the past. This was the loophole mentioned on p. 153.

At a more esoteric level, B. Carr in 1975 invoked the Hawking process (see Chapter 4) to explain the microwave background. We recall that in the

Fig. 8.4. The microwave background spectrum: at the top the ground-based observations; below the observations made from space, based on the work of D. P. Woody and P. L. Richards.

Hawking process a black hole of very small mass can be very hot. The temperature of a black hole in terms of its mass is given by the formula

$$T \simeq 6 \times 10^{-8} \left(\frac{M_\odot}{M} \right) K.$$

Thus a black hole formed by collapse of a massive star ($M > 3 M_\odot$) will be very cold. Hawking had, however, argued that just after the big bang there may be inhomogeneities of matter density which *could* lead to the formation of *mini black holes* – black holes much smaller in mass than $M_\odot$ $= 2 \times 10^{30}$ kg. Carr investigated these primordial black holes in detail and came to the conclusion that provided the mass-spectrum of these black holes is of suitable type they can generate sufficient radiation by evaporation to account for *all* the photons in the microwave background. Carr's mini black holes are as small as 10^{-8} kg in mass and they evaporate in a characteristic time of $\sim 10^{-43}$ seconds!

Any process which aims at explaining the microwave background must quantitatively supply the answer about its present intensity. Why does the background have the energy density of $\sim 10^{-14}$ J/m³? This question is often phrased in the language of the particle physicists in the following way. The observed mean density of matter in the Universe is $\sim 10^{-28}$ kg/m³ and this corresponds to ~ 1 baryon in a volume of 10 cubic metres. The microwave background radiation at the present epoch contains $\sim 5 \times 10^9$ photons in the same volume. Although the number densities of baryons and photons change as the Universe expands, their *ratio* remains unchanged. Thus the photon to baryon ratio is $\sim 10^8$–10^{10} and we have to explain how this large number arises. Even the hot big bang scenario has so far not provided the answer to this question.

Cosmology and Particle Physics

The success of primordial nucleosynthesis and the observation of microwave background have inspired many theoreticians to be more daring and to push their speculations to even earlier epochs than the 1/100th of a second of Table 8.1. Such investigations, first begun by Ya. Zeldovich in the U.S.S.R. in 1965 and by H. Y. Chiu in the U.S.A. in 1966, aim at explaining the abundances of various elementary particles in the Universe. Thus these investigations do for particles what Gamow did for nuclei.

During the 1970s important advances were made by particle physicists towards understanding the overall pattern which relates the various interactions of physics (see Appendix D) to each other. The work of A. Salam, S. Weinberg, and S. L. Glashow showed how we may look upon the weak and the electromagnetic interactions as different aspects of one basic interaction. It is felt by many particle physicists that an enlargement of this pattern will also show how the strong interaction can also be brought into the same fold. The so-called Grand Unified Theory (GUT) epitomizes this goal of unifying these three basic interactions of physics.

One of the key features of GUT is energy. A low energy particle can make a distinction between the three interactions. If we raise the energy high enough – to the theoretically estimated value of 10^{11} eV, the particle ceases to distinguish between the weak and the electromagnetic interaction. At the still higher energy of $\sim 10^{24}$ eV, all three interactions become indistinguishable from one another and GUT shows its full effect.

Now energies $\sim 10^{24}$ eV are beyond the scope of our present high energy accelerators. They were, however, attained by particles in the very, very early stages of the Universe after the big bang. This has prompted a number of exciting theoretical investigations into the early Universe to see if GUT can tell us how particles were formed. At the time of writing this book, no unique answer to this question has emerged and the following is a rather sketchy and incomplete account to give the reader some flavour of the present work.

These investigations proceed along the following line. We recall the dependence of temperature T on time t mentioned before. The constant of proportionality (α in the formula on p. 150) depends on the number of species of particles and their spin states. A photon, for example has two spin states. So does an electron or a positron. A neutrino, on the other hand, has only one spin state. It is a general rule of thermodynamics that if a system of particles is in thermal equilibrium, there is an equipartition of energy between the different states of all the species of particles. The temperature of the entire system therefore depends on what types of particles make up the system. This feature plays a crucial role in the calculation of abundances of different particles.

For the temperature characterizes the kinetic energy of a typical particle and in deciding about survival/creation/annihilation of the particle it is necessary to see how high this energy is *vis-à-vis* the interactions that the particle can have with other particles. Such interactions transform one species of particles into another. Some interactions happen fast and frequently, while others are slow and infrequent. The interactions can be strong, weak, or electromagnetic. These differences show up in the behaviour of the different particles. For example, weakly interacting particles (like the neutrino) dissociate themselves from the family of all particles fairly early in the game when their temperatures are quite high. At the other extreme, the strongly interacting particles (like the neutron or proton) remain hooked to one another for a

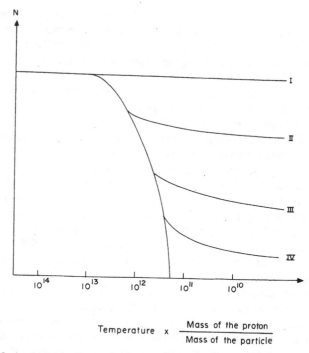

Fig. 8.5. A schematic diagram showing how the numbers of different species of particles in the very early Universe change with falling temperature. The top line (I) shows massless or weakly interacting particles. The next line (II) corresponds to neutral leptons. line III to electrically charged leptons, and the last line (IV) to hadrons which are the least abundant. In general the stronger the interaction the less the abundance. The temperature scale above is modified by multiplication with the ratio of the proton mass to the particle mass.

longer time until the temperature has fallen sufficiently low for them to decouple themselves from one another. In this period their numbers drop appreciably through particle–anti-particle annihilations. The general trend is illustrated in Fig. 8.5.

The detailed calculations involve models of strong interaction built by particle physicists. For example, *quarks* and *gluons*, invented by particle physicists to understand the nature of strongly interacting particles, have to be invoked. These considerations take us even further back in time than in Fig. 8.5. Thus we can ask whether free quarks have survived to this day or whether all of them combined themselves into particles. Another puzzle which needs to be solved is the observed excess of baryons over anti-baryons (i.e. the predominance of matter over anti-matter) in the Universe. If the Universe started existence in a symmetrical manner with equal quantities of matter and anti-matter, how did it come to have the presently observed asymmetrical composition in this respect?

These considerations take us to the epochs when the Universe was only $\sim 10^{-6}$ second old, an epoch when quarks were found in a free state. The average separation between quarks may be as small as $< 10^{-16}$ m. Normally three quarks make up, say, a proton. But with such a sea of quarks it is not possible to argue that any three specific quarks came together to form a proton or other nuclear particle (called a nucleon). However, with the passage of time the temperature drops, the separation increases, and at the age of $\sim 10^{-5}$ second three quarks may collect together in a bag which eventually forms into a nucleon. J. Pati, one of the workers in this area, has labelled this epoch as one of '*quarko synthesis*'. Considerations of energy lead to the conclusion that around this epoch the quarks settled down to masses of the order of 600–700 times the electron mass, which they retain to this day. It is during this period of 10^{-6}–10^{-5} s that the baryon-non-conserving reactions may have taken place, leading to an excess of baryons over anti-baryons. In some alternative models, however, this asymmetry is believed to have taken place even earlier, when the Universe was less than 10^{-30} second old! Some particle physicists and cosmologists advocate the study of the Universe when it was only 10^{-44} s old, in order to understand how gravity is related to the other basic laws of physics. We do not wish to go further into these highly speculative aspects of particle physics and cosmology here.

It is worth pointing out, however, that primordial nucleosynthesis *may* be able to put an upper limit on how many types of neutrinos can exist in nature. Three types are already known: the electron neutrino (v_e); the muon neutrino (v_μ); and the tau neutrino (v_τ). Do more species exist? It appears that the calculated helium abundance increases with the addition of each new species of neutrino. The present calculated abundance of ^{4}He is already high enough so that most workers believe that no new species of neutrinos remain to be discovered.

Conclusion

This concludes our discussion of the big bang and its immediate aftermath. The instant $S = 0$ itself lies outside the purview of physics but this barrier can be approached arbitrarily closely depending on how daring one can be in extrapolating our laboratory physics. To many physicists such investigations seem highly speculative while to some these are worth while, if only because they may throw more light on our

understanding of basic physics. Speculative though they are, most
- physicists would regard such investigations as lying within the domain of
conventional physics.

In the next (and last) chapter we will adopt a somewhat unconventional
attitude and argue that the big bang does not present an insurmountable
barrier. Rather, we show that it may be possible to think of the Universe
before the big bang and that such considerations provide fresh clues to
some of the violent phenomena in the Universe. To be able to discuss this
important issue we have to widen our framework for describing gravity
beyond that of general relativity.

9 The Big Bang Revisited

Twinkle twinkle quasi-star!
Are you near or are you far?

Our discussion of the high energy phenomena in astrophysics has so far been within the framework of Newtonian gravity and Einstein's general relativity. We have seen that where the gravitational effects are weak (in the sense described in Chapter 4) Newtonian gravity provides a simple working theory, whereas the cases of strong gravitational effects are dealt with better by the theory of relativity. For stellar structure and evolution leading to the supernova explosion the Newtonian framework is used, whereas for binary X-ray sources, active galactic nuclei, and for quasars the use of black holes implies recourse to general relativity. In the last chapter on cosmology we saw that the Friedmann models were also based on relativity.

Is general relativity adequate for describing these violent phenomena of strong gravity? The more conventional physicists – and these constitute the majority – would tend to answer this question affirmatively. According to them, astronomy has yet to produce a phenomenon which forces one to consider something new in basic physics. It is true, so they would argue, that many of the events discussed in the earlier chapters require extrapolation from basic physics far beyond the limits which have been tested in the terrestrial laboratory. But, since these extrapolations seem to be working well, where is the need to look for new physics?

There is a small minority of theoreticians who do not take this complacent view. According to them, astronomy has already generated enough problems of a conceptual and observational nature to warrant some rethinking of fundamental physics. Since, as we have seen throughout this book, gravity plays a key role in controlling the violent phenomena in the Universe, much of this rethinking is focussed on the nature of gravity. It will not be possible to do justice to the many thought provoking approaches which exist in this area.* We will concentrate here on one approach with which this author has been personally concerned.

---

* For a review of such approaches the reader is referred to the article *Non-Standard Cosmologies* by J. V. Narlikar and A. K. Kembhavi published in *The Fundamentals of Cosmic Physics* by Gordon and Breach (1980).

Inertia and Cosmology

Let us begin with a striking observational result about the nearest astronomical object, the Earth!

We can measure the Earth's axial rotation in two different ways. Of these, the astronomer's way is to look at the rising and setting of stars and to conclude that the Earth completes one revolution about its axis in one day. This, of course, is the age-old method used by man from primitive days, although at first he interpreted this observation as indicating a cosmos revolving round the fixed Earth.

The second method, which needs some discussion here, makes use of Newton's laws of motion adapted to a rotating frame of reference. When Newton formulated his laws of motion his one major conceptual difficulty was with frames of reference. This difficulty can be understood in the following way. When specifying the motion of a body we need to specify a background reference frame against which the motion is measured. To say that a car is moving north at the speed of 50 m.p.h. is not enough. We have to say what is the reference frame relative to which it is moving. It is usually tacitly assumed that the car is moving relative to the Earth's surface. However, an observer moving in a train keeping pace with the car would find the car *at rest* relative to him. Moreover, the Earth's surface is not at rest relative to, say, the centre of the Sun. So relative to the centre of the Sun the car is not moving at 50 m.p.h., but it is moving at an altogether different and greater speed. Therefore there is no such thing as *the* velocity. Nor is there a unique *acceleration* which can be ascribed to a moving body: we have to specify the frame of reference against which the acceleration is to be measured.

Now consider Newton's second law of motion,

force = mass × acceleration,

telling us that the force acting on a body equals the product of its mass and acceleration. If the same force acts on two bodies one of which is ten times as massive as the other, the less massive body will acquire ten times the acceleration of the more massive one. This result is often stated qualitatively in this way: the more massive body has more inertia and hence it is more difficult to change the state of its motion.

So far so good. But what is the reference frame against which this acceleration is to be measured? Unless this is specified the second law of motion also does not make sense! Newton realized this difficulty and he discussed this problem in terms of an experiment of a rotating water filled bucket. Suppose we have two water filled buckets in the laboratory. One is at rest while the other is set rotating about its axis of symmetry, as in Fig. 9.1. Notice that the surface of the water in the stationary bucket A is

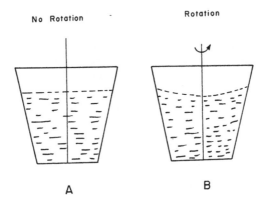

Fig. 9.1. Two water filled buckets A and B shown side by side. The water surface in the stationary bucket (A) is horizontal, while that in the spinning bucket (B) is curved.

horizontal, while that in the rotating bucket B is curved. Thus there is an observable difference in what happens to the water in A and B. How do we explain this difference? Using Newton's law of motion we can argue that water in A has a level surface because it is at rest, while that in B has its surface curved because it is accelerating. However, if we were observing these buckets from a different frame of reference, from the rest frame of B, how would we explain this observation?

A fly sitting on bucket B would conclude that it is at rest while the bucket A is moving. Why then is the water surface in bucket B curved and that in A level?

Clearly the situation of buckets A and B is not symmetrical with regard to Newton's law of motion. Bucket A seems to have special status. In other words, there is a unique reference frame in which Newton's law of motion seems to work – the frame in which bucket A is non-rotating. If we insist on describing the situation from the rest frame of the fly on bucket B, we have to invent fictitious forces to account for the fact that the water surface in B is curved.

Realizing this feature, Newton postulated the so-called *absolute space* in which his laws of motion are valid *without* the need to invent fictitious forces. In the above example bucket A was at rest relative to the absolute space. The ficitious forces like those needed to explain the curvature of the water surface in the rest frame of bucket B are called *inertial forces*.

Experiments such as the one described above can tell us whether a particular frame of reference does or does not correspond to the absolute space. For example, we can enquire whether the frame of reference fixed on the surface of the Earth is rotating relative to the absolute space, and this brings us to our second method of measuring the Earth's rotation.

Of course, in principle, if we place a water filled bucket on the North Pole we should expect to see its water surface curved, because we think that the Earth is rotating about its axis relative to the absolute space. The effect, however, is very small. There is another more noticeable way in which the effect of fictitious inertial forces can be detected. This method uses the *Foucault pendulum.*

A Foucault pendulum is an ordinary pendulum which can oscillate in any vertical plane through its point of suspension. (A clock pendulum is, by contrast, constrained to oscillate in *one* vertical plane.) If the Foucault pendulum is set oscillating in a vertical plane, it is found that gradually its plane of oscillation rotates. This rotation is caused by the inertial forces and their observation therefore establishes the existence of such forces.

Calculations tell us that if the Earth is rotating relative to the absolute space with an angular speed Ω about its polar axis, then at the poles the plane of the Foucault pendulum should also rotate with the angular speed Ω *in the opposite sense.* If the experiment is performed at a lower latitude the angular speed decreases, being zero at the equator. (Mathematically, the angular speed at latitude l is $\Omega \sin l$.)Thus if we measure the speed of rotation of a Foucault pendulum at a place of known latitude we can determine Ω.

The remarkable result that emerges from such a measurement is that Ω turns out to be the same as the Earth's angular speed of rotation relative to distant stars. The result is remarkable because our second method apparently had nothing to do with astronomy! What should one make of such a result? Is it just an accidental coincidence or does it hide some fact of deep significance?

Although Newton's laws were hailed as providing a successful foundation of the science of motion, their foundations troubled a few philosophers of science. In particular, Bishop Berkeley, a contemporary of Newton, criticized the concept of absolute space only a few years after the publication of Newton's *Principia.* It was Ernst Mach (1838–1916), another critic of Newton in the nineteenth century, who first drew attention to the above result of measuring the Earth's rotation speed in two ways. And Mach considered the result to be of deep significance.

According to Mach, the significance of this result lies in the fact that it relates the process of measuring inertia to the background of distant stars. The second law of motion can be looked upon as measuring mass (which is a quantitative estimate of inertia),

$$\text{mass} = \frac{\text{force}}{\text{acceleration}},$$

and the absolute space (in which this law holds) is none other than the frame of reference in which the distant stars are non-rotating. On this basis Mach argued that the property of inertia is somehow related to the background of distant stars.

This rather vague and qualitative concept is known as *Mach's principle*. Its observational basis has, if anything, been strengthened in recent years. Although astronomical observations have shown stars to be slowly rotating around the Galaxy, the background of the distant galaxies does seem to be non-rotating relative to the local absolute space.

The quantitative description of Mach's principle has interested several physicists in the past, including Einstein. In his early days while formulating general relativity Einstein had hoped to incorporate Mach's ideas in his theory of gravity. In this attempt he was ultimately unsuccessful and later came to doubt the validity of Mach's principle. Among the more recent attempts at incorporating Mach's principle into theories of gravity may be mentioned those of D. W. Sciama, R. H. Dicke, D. Lynden-Bell, B. Bertotti, F. Hoyle, and the author.

In our formulation we started with a mathematical formula linking the mass of a typical particle with the presence of the rest of the particles in the Universe. It was therefore a formula giving a direct quantitative expression to Mach's principle. Yet although it started as a theory of inertia, in the end it led to a framework describing gravity, from a wider perspective than general relativity. In the remaining part of this chapter we will outline the basic features of this theory of gravity and its implications for astronomy and cosmology.

Scale Invariance

If the mass of a particle depends on the rest of the Universe, it need not be fixed. As the particle moves across the Universe, its mass may change. This is the first consequence of the Machian theory of gravity described above. What does this result imply for physics in general?

Normally, a physicist expresses any measured quantity in certain basic units. For example, in the International System of units mass is expressed in kilograms, electric charge in coulombs, power in watts, and so on. Not all these units are independent, however. Thus the velocity is expressed in metres per second, of which *metre* is the unit of length and *second* the unit of time. In fact a careful scrutiny will soon confirm that *all* units can be expressed in terms of at most three units; those of length, mass, and time.

However, two fundamental developments in this century have further reduced this number from 3 to 1. The special theory of relativity has established the existence of a fundamental velocity, c, which happens to equal the speed of light in vacuum. The quantum theory has established another fundamental constant, $\hbar$, the quantum of action.* If we set $c = 1$ and $\hbar = 1$, all three basic units of length, mass, and time can be expressed in terms of only one of them. Table 9.1 gives examples of how the various physical quantities can be expresed in terms of M, the unit of mass.

Table 9.1 Physical Quantities in Mass Units $(c = 1, \hbar = 1)$

Quantity	Units in M, L, T	Units in powers of M
Length	L	M^{-1}
Velocity	LT^{-1}	M^0
Force	MLT^{-2}	M^2
Electric charge	$M^{\frac{1}{2}}L^{\frac{3}{2}}T^{-1}$	M^0
Magnetic field	$M^{\frac{1}{2}}L^{-\frac{1}{2}}T^{-1}$	M^2
Angular momentum	ML^2T^{-1}	M^0
Gravitational constant	$M^{-1}L^3T^{-2}$	M^{-2}

Notice how the different units occur as powers of M. Once we fix M all units are fixed in magnitude. We can take, for example, the mass of a proton to be equal to the basic unit M. With this fixation *all* physical quantities are expressible as pure numbers. A few typical examples are given below:

Mass:1 kg. $= 5.98 \times 10^{26}$

Magnetic field:1 T $= 7.86 \times 10^{-20}$

Energy:1 J $= 6.65 \times 10^9$

Gravitational constant:G $= 5.90 \times 10^{-39}$.

Let us now go back to the Machian result that M, the mass of a typical particle, may change with space and time. In that case, the numerical value of any measured quantity will also change, depending on the way it varies with M. For example, the unit of length varies as 1/M. Hence as M increases this unit will decrease and the same segment will have a larger number of the length units than before. The possibility of such changes introduces the interesting question: 'Are the laws of physics invariant under a change of scale?'

Scale invariance of a physical law implies the following property. If the basic unit of measurement changes arbitrarily with space and time, does this information show up in the basic equations of the physical law? If the answer to this question is 'no', then the law is scale invariant. Of the present basic laws, those of electromagnetism describable by Maxwell's equations are scale invariant, and so is the law of propagation of a neutrino. General relativity, on the other hand, is *not* scale invariant. And

---

* This constant is none other than the Planck constant h, encountered in Chapter 4, divided by 2π.

this in turn implies that the equations of this theory cannot successfully incorporate the notion of particle masses varying with space and time. In general relativity particle masses must remain strictly constant. To have a scale invariant theory of gravity we have to go to a wider theoretical framework than general relativity.

Such a framework *is* provided by the Machian theory of gravity. Its basic equations look, at first sight, more complicated than those of general relativity because they permit physics to be discussed in terms of arbitrarily changing basic units. Yet this versatility turns out to be an asset in the long run since it allows us to look at cosmology in a simpler way. We will describe now how this comes about.

Is the Universe Expanding?

In the last chapter we described the red-shift of galaxies, first observed by Hubble, and discussed how this phenomenon can be explained in terms of the expanding Universe. The Friedmann models of the expanding Universe are models of non-Euclidean geometries whose principal feature is contained in the scale factor S. If a light photon of wavelength λ_1 is emitted by a galaxy G at an epoch when the scale factor was S_1 and it is received at the present epoch with the scale factor S_p, its wavelength at reception is red-shifted to

$$\lambda_p = (1 + z)\lambda_1 = \frac{S_p}{S_1}\lambda_1,$$

where z is the observed red-shift. How does this red-shift arise? It arises because the photon travels through the non-Euclidean space-time of the Universe.

Now let us look at the same picture from a different point of view, using the scale-invariant gravity theory described in the last section. It can be shown that the equations of this theory do not require the space-time geometry to be non-Euclidean. A much simpler solution exists wherein the geometry is Euclidean and the Universe is *non*-expanding! This simplicity of geometry is, however, achieved at a cost. The masses of elementary particles are now epoch dependent.

How then is the red-shift of the galaxy G explained? First we note that the mass of any particle, say the electron, was less at the time the light photon left G than the mass of the same particle at the present epoch. Let the two masses be respectively m_1 and m_p with $m_p > m_1$. Now quantum theory tells us that, other things being the same, the wavelength of the spectral line arising from a transition of an atomic electron varies in inverse proportion to the mass of the electron. Therefore if the wavelength of the photon emitted by G in an atomic transition was λ_p, then a photon emitted at the present epoch in a similar atomic transition will have a shorter wavelength λ_1. Thus the observed red-shift z is simply the ratio

$$1 + z = \frac{\lambda_p}{\lambda_1} = \frac{m_p}{m_1}.$$

Note that the wavelength of the photon from G has not become longer during transit: it has stayed the same, at the longer value λ_p it had at origin. The old photon happens to have a longer wavelength than a photon produced now because it originated at an epoch when the particle masses were smaller than at present. The red-shift is therefore direct evidence of the fact that particle masses were smaller in the past.

The two pictures are shown in Fig. 9.2. In (i), describing the standard Friedmann model, the red-shift arises from the expansion of the Universe,

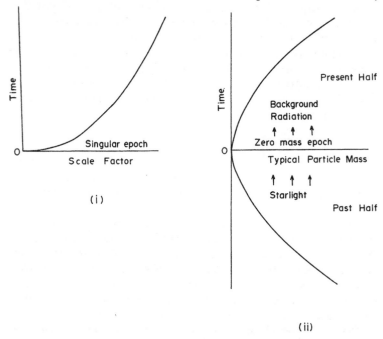

Fig. 9.2. Two ways of describing cosmology. In the standard Friedmann picture, shown in (i) on the left, the scale factor S increases from 0, indicating an expansion of the Universe from a singular state. The particle masses are constant in this picture. In (ii) on the right the Universe is static (i.e. the scale factor is constant) but the masses of particles increase from 0. Unlike (i) this picture has no geometrical singularity at $m = 0$, and it can be continued further back in the past where the particle masses again increase. Hoyle argued that starlight crossing in the direction of the arrows is thermalized and appears as the background radiation.

the masses of particles being constant. In (ii) the Universe is static and Euclidean and the red-shift arises from variable particle masses. The two pictures are partly but not completely equivalent.

The departure between (i) and (ii) is noticeable as we approach the big bang epoch of $S = 0$ in (i). In (ii) this epoch corresponds to the epoch of vanishing mass, $m = 0$. In (i) we have a space-time singularity identified with the origin of the Universe. In (ii) we have no such geometrical singularity; only the masses vanish.

It is here that the advantage of (ii) over (i) begins to show up. Recall that while discussing the big bang in Chapter 8 we commented on the unsatisfactory nature of the way the Universe is suddenly created. This feature is absent in (ii). The Universe can and does exist prior to the zero mass epoch. There is no singularity of geometry preventing us from continuing our investigations *through* the $m = 0$ epoch. We will call the part of space-time preceding this epoch as the 'past half' of the Universe. The part *after* the $m = 0$ epoch will be called the 'present half' of the Universe. In the standard Friedmann description of (i) we have the equivalent of only the present half of (ii).

In fact Fred Hoyle pointed out in 1975 that the epoch of zero mass can play a crucial role in explaining the radiation background observed in microwaves. Any starlight or other form of radiation passing across such an epoch from the past to the present half of the Universe gets efficiently scattered by zero mass particles and is thereby thermalized. This explains why the Universe was dominated by thermalized radiation at high red-shifts. The standard 'hot big bang' picture described in the last chapter did not tell us *why* the Universe was hot to begin with.

We can therefore argue that in the wider framework of scale-invariant gravity there are no space-time singularities but there exist epochs when particle masses vanish. This remark is valid not only for the big bang singularity of the Friedmann models but also for the more general types of singularity in relativistic cosmology, as was shown by Ajit Kembhavi in 1978. The relationship of space-time singularity to a zero mass epoch is discussed further in Box 9.1.

Box 9.1 Zero mass fields and space-time singularities

The geometry of the flat space-time of special relativity is given by the line element

$$ds^2 = dt^2 - dx^2 - dy^2 - dz^2,$$

where (x, y, z) are the Cartesian coordinates and t the time coordinate. A scale transformation changes the element of length ds to

$$d\bar{s} = \Omega\, ds$$

where Ω is known as the *conformal function*.

In the scale-invariant theory of gravity the masses vary with epoch, a typical mass m being given by

$$m = m_p \left(\frac{t}{t_p}\right)^2.$$

The subscript p denotes the present values of m and t. If we make the above scale transformation we must correspondingly change the mass to

$$\tilde{m} = \frac{m}{\Omega}.$$

(Remember: in our system of units mass goes as the reciprocal of length.)

Now Einstein's general relativity assumes that particle masses are constant throughout. Can we therefore obtain the relativistic version of the flat space-time Universe described above? We can, if we are allowed to choose a conformal function Ω which will make $\tilde{m}$ constant.

To achieve this we obviously need $\Omega \propto t^2$. Setting $\Omega = t^2/t_p^2$ we get our new line element as

$$d\tilde{s}^2 = \frac{t^4}{t_p^4}(dt^2 - dx^2 - dy^2 - dz^2).$$

If we make a coordinate transformation

$$\tau = \frac{t^3}{3t_p^2},$$

we get $\tau_p = t_p/3$ and

$$ds^2 = d\tau^2 - \left(\frac{3\tau}{t_p}\right)^{\frac{4}{3}}(dx^2 + dy^2 + dz^2).$$

This is in fact the line element of the type-B Friedmann model described in Chapter 8, a model also known as the Einstein–de Sitter model. If H is the present value of the Hubble constant we have $\tau_p = 2/3H$.

Are we, however, permitted to make this scale transformation? A cardinal rule of the validity of a scale transformation is that the conformal function should not be zero or infinite. We note, in this case, that Ω does vanish at $t = 0$, $\tau = 0$. The transformation from the flat space-time to the relativistic model is therefore invalid at $t = 0$, $\tau = 0$. The penalty paid for violating our cardinal rule is that in the Einstein–de Sitter model we have to contend with a space-time singularity at $t = 0$, $\tau = 0$.

The work of Ajit Kembhavi shows that by and large the space-time singularities in relativity can be traced to the violation of the validity of scale transformations at $m = 0$ epochs.

White Holes

The zero mass epoch occurs in the above cosmological model because, in the Machian description of inertia, the contributions to the mass of a particle from the rest of the particles in the Universe add up to zero. The total mass contribution adds up to zero because some particles contribute positively and others negatively to the inertia. The zero mass epoch is therefore somewhat analogous to the surfaces of zero electric potential amid a distribution of positively and negatively charged particles.

The analogy is not exact but it helps in visualizing the space-time in general as being divided into different regions by 'zero mass surfaces', as shown in Fig. 9.3. In this figure we have taken a general view of space-time – not the simplified homogeneous and isotropic, and thus highly symmetrical, view usually taken in cosmology. This is to show that the Universe may be much more complicated in structure than the impression we have formed with our limited observational means. In any case it is a mystery that the characteristic size of the expanding Universe in the simplified standard cosmological models, a size of the order of ten

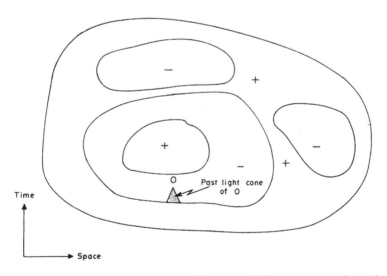

Fig. 9.3. The zero masses surfaces shown in general. They separate regions of positive and negative contributions to masses, shown by + and − signs. An observer like O situated near a zero mass surface may mistakenly think the Universe to be homogeneous and isotropic from his very limited observations confined to his past light cone (shown shaded here). This may well be the case with our present assessment of cosmology!

thousand million light years, should also be the limit up to which our best telescopes can 'see'!

In Fig. 9.4 we examine a typical closed zero mass surface more closely. We consider a bunch of world lines crossing this surface twice, once in the region P_1 and then in the region P_2. How would this situation look within the more limited framework of general relativity?

The relativist will interpret P_1 and P_2 as space-time singularities. His equations therefore do not permit him to cross these singular epochs and he will at best consider the picture in one of its three parts. In part I we have the bunch of particles heading for a space-time singularity. This is the picture of gravitational collapse described in Chapter 3. Although the singularity is reached at P_1, it is hidden from an outside observer O (whose world line is shown in the figure as not crossing the zero mass surface) by a black hole horizon. In region II the bunch of particles appear to erupt from a singularity at P_1 and head towards a singularity at P_2, while in region III there is re-emergence of the particles from a singularity at P_2.

This emergence from a singularity is analogous to the primordial explosion of a big bang Universe. On a limited finite scale such an eruption

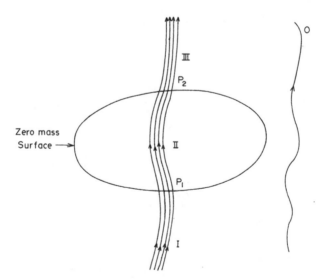

Fig. 9.4. The bunch of world lines intersecting a zero mass surface may be interpreted as giving rise to two pairs of black holes and white holes. The relativistic description of the phenomenon is necessarily limited to one of the three bits, I, II, and III. In I a black hole is formed, in II a white hole gives way to a black hole, while in III a white hole is formed.

is called a *white hole*. In the sense described above a white hole is a time reversed version of a black hole.

In standard general relativity white holes have been relegated to the status of poor relations to black holes. The reasons are briefly as follows. Whereas a black hole is formed from gravitational collapse of a massive object which ends in a singularity, the white hole *begins* in a singularity and ends in the state of an ordinary object. The physicist tends to be prejudiced against an object whose beginning he cannot understand, although the big bang Universe appears to be an exception to this rule! In fact white holes were first proposed as isolated instances of *mini bangs* in regions where the primordial explosion was delayed. The second reason for not taking white holes seriously has been a few calculations showing that they are not stable. These calculations show that, in a very short interval after they erupt, the white holes are smothered by the surrounding medium and they are converted to black holes. In this sense they correspond to region II of Fig. 9.4.

In the wider framework of scale-invariant gravity we see that what is interpreted by O as a black hole in I is causally connected with what is seen in III as a white hole. It is because the relativist tends to see the parts I, II, and III in isolation that he finds that a white hole has a singular beginning, a beginning which cannot be pursued into the past. Unless the full picture I + II + III is taken into account we will not get the proper perspective on white holes. For this reason I think that the calculations leading to the instability of white holes need to be reexamined in the wider framework of scale-invariant gravity.

If we accept that white holes of type III exist, we have an energy machine for generating high energy particles and quanta. For the external observer like O sees the white hole surface approach him with tremendous velocity and hence the quanta emitted from this surface are strongly blue-shifted by the Doppler effect. This blue-shift is so strong in the early stages of eruption that it wins over the large gravitational red-shift arising from the compact distribution of matter in the white hole.

In 1974 K. M. V. Apparao, N. Dadhich, and the author investigated the details of this process. We found that the large blue-shifts can convert an optical photon at source into an X-ray or a γ-ray photon at the observer. This effect, however, does not last long. A typical feature of a white hole is that the energy spectrum of particles and quanta produced by it *softens* rapidly with time, because the blue-shifts do not last long. The characteristic time over which large blue-shifts are seen in a white hole of mass M is

$$\tau = \frac{2GM}{c^3} \simeq 10^{-5} \left(\frac{M}{M_\odot} \right) \text{ seconds.}$$

Thus to sustain a burst of energy for a period of ~ 1 second we need a characteristic mass of $10^5 \, M_\odot$. It is clear that white holes cannot be invoked to explain a sustained production of energy at a steady rate over an extended period of time. They are useful to explain situations where a large amount of energy is produced in a short period. The production of high energy cosmic rays, the bursts of γ-rays or X-rays, or the explosions in the nuclei of galaxies fall into this class. Whatever their conceptual shortcomings within relativity, the white holes hold one advantage over black holes. They are readily and directly observable. Indeed if the theoretical difficulties associated with white holes within the conventional relativity are sorted out, they will turn out to be useful tools for high energy astrophysics.

Anomalous Red-shifts of Quasars

We end this chapter with a brief discussion of a growing set of observations which are proving embarrassing to conventional physics. These observations are about quasars and go under the title of *anomalous red-shifts*.

Basically, the anomaly occurs in the following way. Suppose we see two objects A and B located very near each other, but situated in a remote part of the Universe. According to the expanding Universe hypothesis discussed in Chapter 8, there is a unique relation between the red-shift of an object and its distance from us, the relation originally found by Hubble and known as Hubble's law. Hence if A and B are really close to each other, their red-shifts as measured by us must be equal:

$$z_A = z_B.$$

This criterion has worked well in the case of galaxies. For example, galaxies in a cluster do show very nearly the same red-shifts. Any small differences in their red-shifts can be accounted for by their random motions in clusters.

In the late 1960s, when doubts were being expressed as to whether the red-shifts of quasars are due to the expansion of the Universe, J. N. Bahcall, M. Schmidt, and J. Gunn showed that there are pairs of quasars and galaxies very close to each other with nearly equal red-shifts. This observation was used to argue that since the red-shift of a galaxy is of cosmological origin (i.e. due to the expansion of the Universe) so is that of the quasar near it. More recently, in 1978 A. Stockton examined several parts of the sky and found that there are 13 cases where a galaxy is seen close to a quasar within an angular separation of 45″. And in all 13 cases the quasar red-shift is very close to that of the galaxy.

The astronomer does not have a very accurate distance indicator for

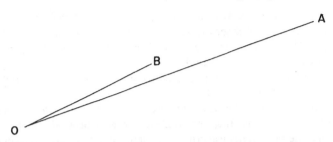

Fig. 9.5. Two objects A and B as seen by the observer O may appear close to one another on the sky if their directions from O are very near each other. It may, however, happen, as shown above, that A is much further away from O than B is.

extragalactic objects. Hence when he sees two objects A and B close to each other on the sky he cannot *ipso facto* argue that A and B are physically near each other. It could happen, for example, that A is much further away than B but their directions happen to be very nearly the same, as shown in Fig. 9.5. How do we know whether this is the case? The question is then settled by statistical arguments. If we know the general population density of, say, objects like A in the sky, we can compute the probability that the direction to A happened by chance to lie close to the direction of B. If the probability works out to be too low, say less than 1 per cent, then there is reason to suspect that A and B are indeed physically near each other.

Arguments such as these were used by Bahcall and others and by Stockton to conclude that the galaxy–quasar pairs observed by them were indeed physically near one another.

However, arguments like these have been used to prove the exact opposite also! In 1971 G. R. Burbidge, E. M. Burbidge, P. Strittmatter, and P. Soloman showed cases of quasar–galaxy pairs of discrepant red-shifts, but with their directions lying close enough to each other to imply that they are physically near each other. Table 9.2 shows the data taken from the 1972 paper of G. R. Burbidge, S. O'Dell, and P. S. Strittmatter.

Table 9.2 Quasar–Galaxy Pairs with Discrepant Red-shifts

Quasar	Galaxy	z_Q	z_G	Angular separation θ in arc minutes
3C232	NGC3067	0.534	0.0050	1.9
3C268.4	NGC4138	1.400	0.0036	2.9
3C275.1	NGC4651	0.557	0.0025	3.5
3C309.1	NGC5832	0.904	0.0020	6.2
3C455	NGC7413	0.543	0.0332	0.4

Since then, thanks largely to the zeal shown by H. C. Arp from the Hale Observatories, the number of such cases has grown rapidly. In a typical case, a quasar Q lies close to a galaxy G with their red-shifts z_Q and z_G substantially different as in Table 9.2. Even Stockton's data, although it was biased in favour of low red-shift quasars, had 12 such cases of discrepant red-shifts!

Whenever such discrepant cases are found, the conventional astronomers try to dismiss them as accidental coincidences, although these cases use the same statistical criteria to establish the physical nearness of quasar and galaxy as Stockton and others did.

The issue would have become bogged down in statistical wrangles, had there not emerged a few other significant features from the data on discrepant red-shifts. These features, briefly summarized below, are hard to explain on the basis of a chance coincidence of directions of quasar and galaxy.

In Figure 9.6 is plotted the angular separation of the quasar–galaxy pairs of Table 9.2 against the red-shift of the galaxy, on a logarithmic scale.

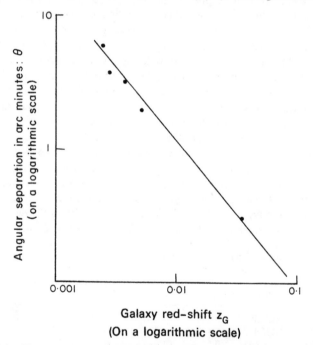

Fig. 9.6. The correlation between the angular separation θ between the quasar–galaxy pair and the red-shift z_G of the galaxy is quite obvious in the above plot of log θ against log z_G.

The points fall close to a straight line of slope -1, indicating that the separation is inversely proportional to the red-shift of the galaxy. This behaviour seems to hold out (although with larger scatter) for about 120 such pairs known at the time of writing this book. Since the galaxy red-shift is a reliable indicator of distance we can argue that the angular separation falls in inverse proportion to the distance of the quasar–galaxy pair from us.

In Fig. 9.7 we see a typical example of two other features which have been found for a number of quasar–galaxy associations. Notice how quasars are aligned on opposite sides of the galaxy in the middle. Also, there are many quasars associated with the galaxy but those aligned across have nearly the same red-shifts.

In a chance distribution we do not expect there to be any correlation between quasar–galaxy separation and galaxy red-shift. It is impossible to say why quasars should be seen aligned across the galaxy if they are there by chance. It is also *impossible* to account for the bunching of red-shifts of the quasars on the hypothesis that they are of cosmological origin. The conclusion seems inescapable that high red-shift quasars are associated with low red-shift galaxies.

Could a substantial part of the red-shift of a quasar be of non-cosmological origin? If z_Q is the quasar red-shift and if z_G is the red-shift of

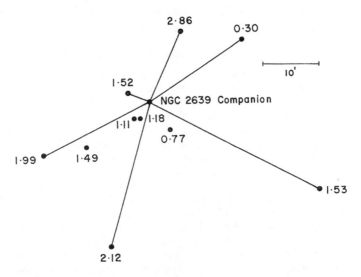

Fig. 9.7. Example of alignment and red-shift bunching for quasars near the galaxy NGC 2639. (Adapted from the work of H. C. Arp.)

the galaxy associated with the quasar, then we can assert that z_G is also the cosmological red-shift of the quasar. Since z_G is in general less than z_Q, there is an *anomalous* component z_A of the quasar red-shift which is given by the relation

$$1 + z_A = \frac{1 + z_Q}{1 + z_G}.$$

Thus if $z_G = 0.05$ and $z_Q = 1.1$, then $z_A = 1$. What could z_A be due to?

At first sight the Doppler effect seems a good candidate for explaining the anomalous red-shift. If the quasars are ejected from the nucleus of the galaxy in a violent explosion it may have a large enough velocity to give $z_A = 1$ in the above example. Alignments seen in Fig. 9.7 could be explained by the quasars being thrown in opposite directions.

However, here comes the main difficulty! If quasars are ejected in all directions from a galaxy then just as some are receding from us (in order to explain the large red-shifts z_A), there should also be some approaching us. These approaching quasars should show large *blue-shifts*. As yet *no* blue-shift, large or small, has been observed in any quasar.

A second way of getting a red-shift is through a strong gravitational field. The quasar, because of its compactness, may have strong gravitational red-shifts. There are, however, problems of an observational and theoretical nature which make it difficult to assume that the red-shift could be due to a strong surface or interior gravity of the quasar. These objections are briefly discussed in Box 9.2.

Box 9.2 Why the anomalous red-shift cannot be gravitational

Gravitational red-shift due to a massive compact object like the white dwarf or the neutron star is well known to the astronomer. It is however quite small; even for a neutron star it is hardly more than $z \simeq 0.3$. The quasars show red-shifts as high as $z = 2$–3. Can these red-shifts be of gravitational origin?

In 1964 H. Bondi showed that a massive object in equilibrium, made of matter obeying normal equations of state (i.e. physical behaviour found in usual laboratory investigations), cannot have a gravitational red-shift exceeding $z \simeq 0.6$. On the observational side, the spectroscopic analysis of the first two quasars, 3C 273 and 3C 48, by M. Schmidt and J. L. Greenstein showed in 1964 that even though these quasars have low red-shifts (0.158 and 0.357 respectively) their narrow emission lines place severe constraints on the models designed to explain their red-shifts as being of gravitational origin.

In 1967 F. Hoyle and W. A. Fowler found a loophole in the above theoretical and observational difficulties. Both difficulties arose, they pointed out, because it was assumed that the light from the quasar comes from its surface. They suggested that larger red-shifts can arise if the light originates in the *central* region of the quasar.

Can light from the centre get out without substantial absorption? Can it have red-shifts high enough, e.g. $z \gtrsim 2$? Can the spectroscopic features of quasars be explained on the hypothesis of a gravitational red-shift? Finally are all these conditions satisfiable without too esoteric a form of matter?

In the mid-1970s P. K. Das investigated these problems at length and answered all the above questions in the affirmative. Nevertheless quasars in these models have to be very massive, even more massive than our Galaxy. It is not clear whether one can accommodate so many of these quasars within a distance of, say, a few million light years, without affecting the dynamical behaviour of nearby galaxies. Thus the collection of a large number of quasars near a galaxy, as found by Arp, would be hard to account for as a dynamically stable and viable system.

Even if these objections were surmounted, it is difficult to say why alignments, and the correlation of the type shown in Fig. 9.6 should be found in a picture based on the gravitational red-shift. Indeed, the gravitational red-shift may at best account for the large red-shift of an isolated quasar, but it *cannot* tell us why there should be any kind of association between a quasar and a galaxy.

A Fourth Kind of Red-shift?

With the cosmological, Doppler, and the gravitational red-shifts ruled out, we are forced to think of some new interpretation for z_A if the data on the anomalous red-shifts of quasars are to be taken seriously. Such an interpretation is readily found in the scale-invariant gravity theory described earlier in this chapter. The central idea in this interpretation is illustrated in Fig. 9.8.

Here we go back to our description of cosmology with variable particle masses. In a completely homogeneous and isotropic Universe the zero mass epoch is a *hyperplane* in space and time. In Fig. 9.8 we consider the consequence of having a kink in such a plane. We also see three world lines in Fig. 9.8. The line O is the world line of an observer (like us!), the line G is the world line of a galaxy, while the line Q is the world line of an associated quasar showing anomalous red-shift. The lines O and G intersect the zero mass surface, shown by a thick line, at the epoch $t = 0$ while the line Q, passing through the kink, meets it at a *later* epoch $t = t_1$.

Suppose the observer receives the light photons from Q and G (which are near each other) at an epoch t_p. The photons left Q and G at the epoch t_2 shown in the figure. What red-shifts will be seen by the observer O?

The answer to this question depends on the particle masses at O (at epoch t_p) and at Q and G (at epoch t_2). The theory tells us that the masses vary in proportion to the *square* of the time elapsed since the zero mass surface was crossed. Denoting the respective masses in O, Q, and G at the

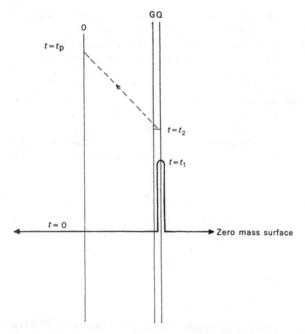

Fig. 9.8. The space-time diagram showing a kink in the zero mass surface. The details are explained in the text.

above mentioned epochs we therefore get

$$m_O \propto t_p^2, \; m_Q \propto (t_2 - t_1)^2, \; m_G \propto t_2^2,$$

the constant of proportionality being the same in all three cases. By our earlier rule discussed in this chapter the red-shifts z_Q and z_G are given by the mass ratios. Thus

$$1 + z_Q = \frac{m_O}{m_Q} = \frac{t_p^2}{(t_2 - t_1)^2}$$

$$1 + z_G = \frac{m_O}{m_G} = \frac{t_p^2}{t_2^2}.$$

Obviously, $z_Q > z_G$ as required.

Thus even though the quasar and the galaxy are physically near each other, the quasar has the larger red-shift because it happens to cross the zero mass surface at a later epoch.

What is the physical interpretation of this picture? The kink in the zero mass surface amounts to an explosion in the nucleus of the galaxy at the

Fig. 1.9. The object shown in the centre is associated with the radio source Cygnus A. Earlier it was argued that this object describes a collision of galaxies. Now the collision hypothesis has been found to be untenable. (Photograph from Palomar Observatory, California Institute of Technology.)

Fig. 5.1. The Great Nebula in Orion which is believed to contain many young pre-main-sequence stars, which emit infrared radiation. The Nebula is probably one of the many locations in our Galaxy where new stars are being born. (Photograph from Palomar Observatory, California Institute of Technology.)

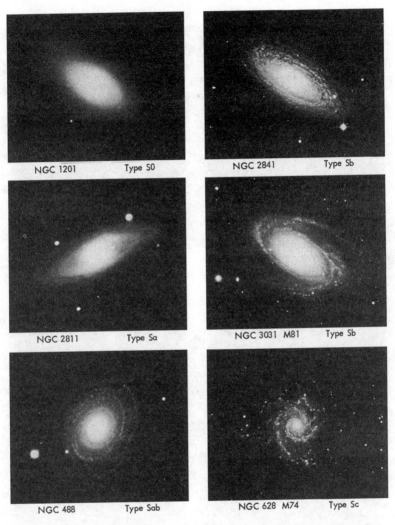

NGC 1201   Type S0

NGC 2841   Type Sb

NGC 2811   Type Sa

NGC 3031 M81   Type Sb

NGC 488   Type Sab

NGC 628 M74   Type Sc

Fig. 7.1. Spiral galaxies of different types. (Photograph from Palomar Observatory, California Institute of Technology.)

Fig. 7.2. The central spherical region of M 87. For four views of the jet in M 87 see Fig. 1.8. (Photograph from Palomar Observatory, California Institute of Technology.)

epoch t_1, an explosion which ejected a newly born quasar. (See the previous section on white holes where the zero mass surfaces are likened to mini explosions.) At birth the particles in the quasar had zero mass, but the mass began to increase with time. However, a typical sub-atomic particle in the quasar would always be less massive than its counterpart in the parent galaxy. It is this difference which accounts for the anomalous red-shift of the quasar.

The red-shift bunching effect is easy to explain under this hypothesis. If two quasars are fired in opposite directions at a single explosion at the epoch t_1, then they will both show the same red-shift, since the red-shift depends only on the epoch of creation. The alignment is explained by the fact that to conserve the overall momentum the quasars must be fired in opposite directions. In Fig. 9.7 multiple explosions have taken place at different epochs.

But what about the blue-shift effect noted earlier? Computer calculations by P. K. Das and the author show that the quasar speeds drop sharply from the near-light speed of ejection and hence the Doppler effect is negligible. Therefore blue-shifts are not expected in this picture. The dropping of speed arises from the circumstance that the quasar mass grows in time. And this effect helps the galaxy in trapping the quasar by its strong gravity. Still some quasars may escape altogether! Thus it is possible to argue in this picture that *all* quasars are born in explosions in galactic nuclei, and the isolated quasars are those which have escaped the gravitational confinement of the parent galaxy.

The quasars which are trapped by their parent galaxies will orbit around the galactic nuclei in highly eccentric orbits of steadily diminishing size. It is these quasars which are picked up by Arp and others in their search for quasar–galaxy associations. The characteristic distance up to which the gravitational influence of a galaxy of mass M extends in a Universe of Hubble constant H is given by

$$d = \left(\frac{GM}{H^2}\right)^{\frac{1}{3}} \simeq 1.3 \times 10^6 \left\{\frac{M}{10^{11} \, M_\odot}\right\}^{\frac{1}{3}} \text{ light years.}$$

Since the *physical* separation between the quasar and the galaxy in an association will be of this order, the *angular* separation between the two would drop off in inverse proportion to their distance from us. Thus we see why in Fig. 9.6 there is an inverse correlation between the angular separation and z_G.

Epilogue

This concludes our discussion of a somewhat unconventional set of topics. We began with an important observation relating to the Earth's rotation, one which led in the last century to Mach's principle. In our attempts to incorporate Mach's principle into a theory of gravity we were led to a framework wider than general relativity. For most practical purposes this framework of scale-invariant gravity is equivalent to general relativity. There are exceptional situations, however, where the advantages of going over to a wider framework than relativity begin to show up.

In particular, we gain a better understanding of the big bang singularity. We can get some insight into how the Universe may in fact be a much bigger system than what we imagine through the present observations and through the standard Friedmann models. The origin of the microwave background, the link between black holes and white holes, and the understanding of the curious phenomenon of the anomalous red-shifts of quasars follow naturally out of such a theory.

In the last analysis, however, a radically new theory is welcomed only if it proves to be indispensable. Although we have tried to make such a case for this theory, the conservative sceptics may want to wait for more observations to come in the future, observations which will settle once and for all the question whether the quasars are far or near.

Appendix A

The Electromagnetic Wave and the Photon

In the 1860s James Clerk Maxwell formulated the basic equations of electricity and magnetism. These equations describe how the electric and magnetic disturbances (known technically as *fields*) change across space and time as a result of the distribution of electric charges and currents in space. One important conclusion to emerge from Maxwell's theory was that the electromagnetic disturbances propagate across space as waves.

The electromagnetic wave is a *transverse wave*, as shown in Fig. A.1. That is, the disturbance can be described by electric field and magnetic field in perpendicular directions in a plane which itself is perpendicular to the direction of propagation. The wave shown in Fig. A.1 is *plane polarized*, that is, the electric field is always in one plane and the magnetic field always in another plane (both planes being at right angles to each other). In an *unpolarized* wave the directions of electric and magnetic fields keep changing at random.

In this figure we also see the wave pattern clearly: the magnitude of the electric (and magnetic) field goes up and down regularly. If we watch the wave as it passes one point in space, we will find the fields repeat their pattern (i.e. go from maximum to minimum and back to maximum) with characteristic periodicity. We denote by v the *frequency* of the wave. That is, the pattern is repeated v times per second. In the same way, if we study the form of the wave across space at a given time, its *wavelength* λ describes the characteristic distance over which the same pattern is repeated (see Fig. A.1).

It is clear that in one second v waveforms (i.e. characteristic patterns) pass a given point, each form being of length λ. Therefore the speed of the wave is $v \times \lambda$. An important result to emerge from Maxwell's equations was that in vacuum

$$v \times \lambda = c,$$

that is, the wave propagates with the *speed of light*.

Indeed Maxwell's equations resolved the mystery of the nature of light. Light is an electromagnetic wave. The different colours in which sunlight is split are simply electromagnetic waves of varying wavelengths, as shown in

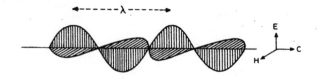

E : electric field
H : magnetic field

Fig. A.1

Table A.1 below. Here the wavelength is expressed in the unit known as the Ångstrom:

1 Ångstrom = 10^{-10} metre.

Apart from these colour bands, the light from an astronomical system, when analysed by wavelengths in a spectrum, also shows dark and bright *lines* at specific wavelengths. The dark lines are called *absorption* lines and the bright lines *emission* lines. These lines appear because of the interaction of light with electrons in atoms and molecules. The interaction can result in the absorption or the emission of light by these electrons. The electrons in atoms or molecules, however, cannot lose or gain energies in arbitrary amounts. The rules of quantum theory fix these amounts to a set of discrete quantities. Consequently, only discrete amounts of light energy are absorbed or emitted in these transitions of electrons.

One of the important developments of early quantum theory was the appreciation of the fact by Max Planck that light not only travels as a wave, but at small levels is concentrated in tiny packets of energy called *photons*. The energy carried by a photon of frequency v is hv, where h is a

Table A.1 Colours and Wavelengths

Colour	Wavelength range in Angstroms
Violet/Indigo	3900–4550
Blue	4550–4920
Green	4920–5770
Yellow	5770–5970
Orange	5970–6220
Red	6220–7700

universal constant called *Planck's constant*. Thus the light emitted or absorbed by an electron in atomic or molecular transitions appears as a photon of specific frequency and wavelength. Therefore an absorption line in the spectrum implies absorption of a photon, while an emission line corresponds to the emission of a photon.

This brings us finally to the question of the different forms of electromagnetic waves. For it is clear that the wavelength of an electromagnetic wave need not be limited to the range 3900 Å to 7700 Å. There may be photons of longer or shorter wavelengths than these also. The wavelengths of these other forms of electromagnetic wave are given in Table A.2 below.

Table A.2 Different Forms of Electromagnetic Waves

Form type	Approximate wavelength range in metres
Radio	$> 10^{-1}$
Microwave and Millimetre-wave	$3 \times 10^{-4} - 10^{-1}$
Infrared	$8 \times 10^{-7} - 3 \times 10^{-4}$
Visible light	$4 \times 10^{-7} - 8 \times 10^{-7}$
Ultraviolet	$3 \times 10^{-8} - 4 \times 10^{-7}$
X-rays	$3 \times 10^{-12} - 3 \times 10^{-8}$
γ-rays	$< 3 \times 10^{-12}$

At very small wavelengths, e.g. for X-rays and γ-rays, it is more convenient to use the energy rather than the wavelength to specify the photon. Using the relation between frequency and wavelength we get the energy of a photon of wavelength λ as

$$E = h\nu = \frac{hc}{\lambda}.$$

The wavelength range $3 \times 10^{-12} - 3 \times 10^{-8}$ m for X-rays from Table A.2 now corresponds to the energy range 0.125 keV–0.4 MeV, while the range of energy of γ-ray photons begins at $\sim$ 0.4 MeV. Here the energy unit used is the *electron-volt* (eV), which equals 1.602×10^{-19} J, and 1 keV = 1000 eV, 1 MeV = 10^6 eV.

Appendix B
The Special Theory of Relativity

Until the end of the nineteenth century physicists had taken for granted the concepts of absolute space and absolute time. In Fig. B.1 we see the space-time diagram of Newtonian physics. P and Q are two events characterized by their location in space by the Cartesian coordinates (x_P, y_P, z_P) and (x_Q, y_Q, z_Q), and by their position on the time axis by the instants t_P and t_Q. As shown in the figure, the event Q occurs later than the event P. *Any* observer would be able to state that the time duration between P and Q is $t_Q - t_P$ and the spatial separation of these events is

$$r_{PQ} = \sqrt{\{(x_P - x_Q)^2 + (y_P - y_Q)^2 + (z_P - z_Q)^2\}}.$$

Although all observers in Newtonian physics had the same rules for measuring time intervals and spatial separations, the laws of motion did single out a special class of observers, the so-called *inertial observers*. These were observers moving under no forces who hence had uniform velocities

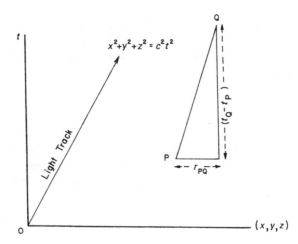

Fig. B.1

relative to one another. The laws of motion had the same form (i.e. they are invariant) as seen by all inertial observers. (In Chapter 9 we discussed how these laws are modified for *non*-inertial observers.)

It was implicitly assumed that all laws of physics should be invariant for inertial observers. The conflict began to appear, however, with the electromagnetic theory. Maxwell's equations (cf. Appendix A) predicted an invariant speed of propagation for the electromagnetic waves, the so-called speed of light c. If any two inertial observers in relative motion with respect to each other are to have the *same* formal structure of Maxwell's equations then they must measure the *same* speed for light. Yet how could two such observers find the same speed for a light beam when they themselves are moving relative to each other? If one observer A measures the light speed as c and the other B is moving towards the beam of light with speed v relative to A then is B not expected to find the speed of light to be $c + v$? Clearly something is wrong somewhere.

On the experimental side, the Michelson–Morley experiment of 1887 had already given indication of there being something wrong somewhere. Attempts to measure the speed of the Earth's rotating surface relative to the hypothetical medium 'aether' in which light was supposed to propagate did not show any such relative motion. Again the underlying calculations made use of velocity addition rules based on Newtonian dynamics.

In 1905 Albert Einstein re-examined the concept of space-time measurements in the light of the expected invariance of Maxwell's equations for inertial observers. And out of this re-examination the special theory of relativity was born. The basic ideas of this theory can be described as follows.

Suppose, in the space-time diagram of Fig. B.1, rays of light leave in all directions from the origin at $t = 0$. Then, if the coordinates of Fig. B.1 are those used by an inertial observer, he will find the light ray at (x, y, z) at time t, where

$$x^2 + y^2 + z^2 - c^2 t^2 = 0,$$

c being the invariant speed of light. Another inertial observer, using coordinates x', y', z', t' with the same origin for space and time, would similarly conclude

$$x'^2 + y'^2 + z'^2 - c^2 t'^2 = 0.$$

Einstein argued that any rule connecting (x, y, z, t) to (x', y', z', t') must take these two results into account. Now for observers in relative motion with respect to each other, these two relations are *impossible* to satisfy if we insist on the Newtonian precept of a universal absolute time. In other words,

$$t \neq t'.$$

Going back to our original example, neither r_{PQ} nor $t_Q - t_P$ have an invariant status for the two events P and Q. Different inertial observers will get different answers for these two measurements. However, the new rules for connecting (x, y, z, t) to (x', y', z', t') advocated by Einstein do keep the quantity

$$s_{PQ}^2 = c^2(t_Q - t_P)^2 - r_{PQ}^2$$

invariant for *all* inertial observers. These rules of connecting the measurements of space and time by two inertial observers are known as the *Lorentz transformation* (first given by Lorentz in another context).

Let us now look at some of the consequences of this assumption.

Suppose an observer S uses the (x, y, z, t) coordinates, which makes him record the events of Fig. B.1 at the instants t_P and t_Q. Let another observer S' move uniformly with speed v relative to S so that he finds himself at the locations of P and Q. That is, he covers the distance r_{PQ} in the time interval $t_Q - t_P$ as seen by S. Therefore

$$r_{PQ} = v(t_Q - t_P).$$

Notice that the events P and Q take place in the *rest frame* of S' so that the spatial displacement of P and Q as measured in the rest frame of S' is zero while the time interval is $t'_Q - t'_P$, say. Accordingly the invariant quantity s_{PQ}^2, as measured by S', is simply

$$s_{PQ}^2 = c^2(t'_Q - t'_P)^2.$$

This must be the same as that measured by S, so that

$$c^2(t'_Q - t'_P)^2 = c^2(t_Q - t_P)^2 - r_{PQ}^2,$$

which gives

$$t'_Q - t'_P = (t_Q - t_P)\sqrt{\left(1 - \frac{v^2}{c^2}\right)}.$$

In other words, if S' is carrying a clock with him, the time interval elapsed in it between P and Q is *shorter* than the difference in the times recorded by two clocks at rest relative to S – one at the spatial location of P and the other at the spatial location of Q – when S' flashes past them. This increase in time interval of fixed clocks relative to a moving clock is known as *time dilatation* and it is usually denoted by the factor

$$\gamma = \frac{1}{\sqrt{\left(1 - \dfrac{v^2}{c^2}\right)}}$$

This dilatation factor γ plays a key role in special relativity. It is because of this that the law of addition of velocities changes from the simple Newtonian one described in Chapter 2 to a more complicated one. But the result is that light always has the same speed c relative to all inertial observers.

Because of these changes in space-time measurements Einstein had to revise the laws of motion. An important difference from the Newtonian concepts is seen in the behaviour of *mass*. If the mass of a body at rest relative to an observer S is m_0, then its mass when in motion is larger than m_0. In our above example if the body is carried by S' its mass *as measured by S* will be

$$m = m_0 \gamma.$$

The observer S can ascribe this increase in mass of the body to its motion, to its kinetic energy. In this way Einstein arrived at his famous formula connecting mass to energy:

$$E = mc^2.$$

A moving mass has larger energy than mass at rest. The difference, the *kinetic energy*, is given by

$$E - E_0 = (m - m_0)c^2 = (\gamma - 1)m_0 c^2.$$

For v small compared to c we recover the familiar Newtonian expression for kinetic energy:

$$E - E_0 \simeq \frac{1}{2} m_0 v^2.$$

This result is characteristic of the way Newtonian mechanics follows from relativistic mechanics. For v small compared to c the two theories agree. For v close to c the distinguishing factor γ becomes large compared to 1 and the two theories differ significantly.

The invariant surface given by the equation

$$x^2 + y^2 + z^2 - c^2 t^2 = 0$$

giving the tracks of light rays emitted at the space-time origin is called the *light cone*. The velocity vector of any material particle (of non-zero rest mass m_o) must lie inside this cone, indicating that $v < c$. Particles with $v = c$ lie *on* the light cone and must have zero rest mass. By firing particles with speeds close to the speed of light in modern high energy particle accelerators physicists attempt to study the behaviour of elementary particles at very high energies. For example, in a typical accelerator experiment a γ-factor as high as 1000 can be achieved. By contrast, in the very high energy cosmic rays $\gamma = 10^{12}$ is found for protons. The origin of these very high energy cosmic ray particles is still a mystery.

Nevertheless it was through cosmic ray particles that the time-dilatation effects were first noticed. The particle called *muon* decays normally in a characteristic time of $\sim 2 \times 10^{-6}$ seconds in its rest frame. To an observer on the Earth a cosmic ray muon coming with large speed seems to last for $2 \times 10^{-6} \gamma$ seconds, where γ is large. In this way muons have been known to survive for as long as $\sim 10^{-4}$ seconds (with $\gamma \simeq 50$).

Appendix C
The Doppler Effect

The Doppler effect describes the change in the observed frequency of a wave arising from a relative motion between the source and the observer. We derive the result for light waves.

Imagine a source of light S moving radially away from an observer O with velocity v. Suppose the source sends waves of frequency v in its rest frame. Thus if it sends one wavecrest at time $t = 0$ (say), the next wavecrest will leave S at the time $t = 1/v$. What is the time gap between the arrivals of these two wavecrests at O?

Had S remained stationary, this gap would have been simply $1/v$. However, since S is moving, by the time it emitted the second wavecrest it had moved further away from O by the distance $(1/v) \times v$. The second wavecrest will therefore have to cover this additional distance before it reaches O, over and above the distance that the first wavecrest covered. Since the wavecrest travels at the speed of light c, this additional distance will be covered by it in time

$$\tau = \frac{v}{vc}.$$

That is, instead of arriving after the time gap of $1/v$, the second wavecrest will arrive after a time gap of

$$\frac{1}{v} + \tau = \frac{1}{v}\left(1 + \frac{v}{c}\right).$$

Hence O will observe the wavecrests as arriving with a frequency

$$v' = \frac{v}{1 + \dfrac{v}{c}}.$$

So far we have assumed the time coordinate to be the same for O and S. However, since S is moving relative to O, special relativity (cf. Appendix B) introduces an additional factor into this calculation: to O the clock of S appears to move more slowly by the factor

$$\frac{1}{\gamma} = \sqrt{\left(1 - \frac{v^2}{c^2}\right)}.$$

Hence the frequency observed by O is not v' but

$$\tilde{v} = \frac{v'}{\gamma} = \sqrt{\left(\frac{1 - \frac{v}{c}}{1 + \frac{v}{c}}\right)} \, v.$$

The corresponding wavelengths are given by the ratio

$$\frac{\tilde{\lambda}}{\lambda} = \sqrt{\left(\frac{1 + \frac{v}{c}}{1 - \frac{v}{c}}\right)}.$$

The red-shift z is given by

$$z = \frac{\tilde{\lambda}}{\lambda} - 1 = \sqrt{\left(\frac{1 + \frac{v}{c}}{1 - \frac{v}{c}}\right)} - 1.$$

For example, $z = 2$ gives $v = 4c/5$. The closer v approaches c, the larger is z. As $v \to c$, $z \to \infty$.

If S were approaching O, the waves would be *blue-shifted*. In general if θ is the angle made by the direction of motion of S with the radial direction from O to S, the wavelengths are related by the formula

$$\frac{\tilde{\lambda}}{\lambda} = \frac{1 + \frac{v}{c} \cos \theta}{\sqrt{\left(1 - \frac{v^2}{c^2}\right)}}.$$

Appendix D

Sub-atomic Particles

In the mid-1930s the standard picture of the atom was formed in terms of a nucleus surrounded by orbiting electrons. A typical atom has Z electrons orbiting a nucleus containing Z protons and $A - Z$ neutrons. The number Z is called the charge number and the number A the mass number. It is customary to specify an atomic nucleus X of mass number A by ^{A}X. The protons are positively charged, each with e units of charge, while the electrons are negatively charged, each with $-e$ units of charge (the magnitude of e in electrostatic units is 4.8×10^{-10}). The atom as a whole is electrically neutral. The protons and neutrons are referred to as nucleons.

The chemical properties of the atom are determined by its charge number Z and its electron structure and these are well understood in terms of the quantum mechanical behaviour of electrically charged particles. Nuclei with the same Z but different A are called *isotopes*. The structure of the nucleus is, however, less easily understood.

For example, how could the nucleus contain, within a characteristic size of $\sim 10^{-14}$ m, so many protons in spite of their mutual electrostatic repulsion? Clearly, the protons and the neutrons in the nucleus are bound by an attractive force considerably stronger than the repulsive electrostatic force. Moreover, this force cannot extend well beyond nuclear dimensions, as the experiments in nuclear physics clearly show.

The attractive force between the protons and the neutrons in the nucleus means, in general, that the nucleus is a tightly bound system. The work needed to tear apart a nucleus is called its *binding energy*. The binding energy gives some indication of how stable and well knit the nucleus is. For example, the helium nucleus is a stable one. A bigger nucleus often breaks apart emitting smaller units in the form of helium nuclei – often called *alpha particles*.

Broadly speaking as we go on adding neutrons and protons together to build bigger nuclei, two opposite effects are called into play. On the one hand the attractive nuclear force increases with the number of protons and neutrons. At the same time, too big a nucleus cannot be held together because of the short range of the nuclear force. The limit of stability is reached for nuclei of the iron group which are most tightly bound. Beyond that, nuclei begin to get too unwieldy to be held together by the nuclear force.

In general it is possible to extract energy by *nuclear fusion* up to the iron group of nuclei. Thus bringing together four ^{1}H-nuclei to make a ^{4}He nucleus releases energy. However, a very big nucleus may release energy by breaking up. This process, known as *nuclear fission*, releases energy in the atomic bomb through the breakup of the Uranium nucleus ^{235}U.

This short range interaction within the nucleus is called the *strong* interaction ('strong' *vis-a-vis* the electrical interaction). Earlier it was believed that this interaction was limited to protons and neutrons and was conveyed by particles called the *pions* which are neutral or charged particles. Certainly the electrons seemed to be immune from this interaction.

Gradually, further experiments revealed more and more elementary particles and they became classified as *hadrons* (heavy particles) and *leptons* (light particles). The leptons, which include the electron, the muon, and the neutrinos, are not affected by the strong interaction. The hadrons, of which the early members are the neutron and the proton, do take part in the strong interaction.

The number of hadrons also grew rapidly as experimental techniques became more sophisticated. In due course it became necessary to drop the qualifying word 'elementary' to describe these particles. For they too show internal structure, being made of sub-units. These sub-units are known as *quarks*. As yet, free quarks have not been discovered: indeed there are theories which claim that quarks can only appear in combination, as particles. The quarks may be held together by binding agents called *gluons*.

The proton is about 1836 times as massive as an electron. The neutron is slightly more massive. There are other particles about as massive as the proton and the neutron, some even more massive. These are all called *baryons*. The pions are about 270 times as massive as the electron and they belong to the family of *mesons*. Both baryons and mesons are believed to be made from quarks. In the original formulation of Gellmann and Zweig, three quarks were needed to form a proton or a neutron and two to form a meson. Now, however, it is felt that more quarks with other characteristics (e.g. charm, flavour . . .) are needed to account for all the observed properties of sub-atomic particles.

There is, however, another interaction known as the *weak interaction* which also seems to govern the behaviour of leptons and hadrons. It was first discovered through the decay of the neutron (a process known as the *beta-decay*) into a proton, an electron, and an anti-neutrino. The interaction, as its name implies, is considerably weaker than the electrical interaction. Yet it plays a crucial role in astrophysics, as we saw in Chapter 5.

Recent developments have gone a long way towards establishing a link between the electromagnetic and the weak interaction. The work of A. Salam, S. L. Glashow, and S. Weinberg has yielded a *unified gauge theory* which combines these two interactions in a single framework. Now physicists are hopeful of some day achieving a *grand unified theory* which also includes the strong interaction in this scheme.

Finally, a word about *anti-particles*. In the late 1920s the work by P. A. M. Dirac led to the result that corresponding to electrons there should exist particles of the same mass but equal and opposite charge. These particles, now known as *positrons*, are in a certain sense the opposites of electrons. A pair made up of an electron and a positron cannot exist for long; the two annihilate each other and produce radiation. Later it became clear that this property of having opposite types was shared by many particles. Thus proton and neutron have *anti*-particles called the anti-proton and the anti-neutron.

This anti-matter as a rule cannot stably coexist with matter and this has led to the speculation as to whether our Universe is entirely made of matter. Certainly if they are situated well apart, clumps of matter and anti-matter can coexist, and it would be difficult to tell from far apart whether a galaxy is made of matter or anti-matter. For both matter and anti-matter have a symmetrical behaviour with respect to light, the sole means at present of studying the remote parts of the Universe. Theoreticians like H. Alfven have advocated, on grounds of symmetry, that the Universe should contain equal amounts of matter and anti-matter. Nevertheless most theoreticians are finding it difficult to cook up reasonable scenarios in which a Universe born in a big bang got separated into different sections, one being matter dominated and the other anti-matter dominated.

Appendix E

Types of Electromagnetic Radiation

Astrophysicists receive electromagnetic radiation in many different forms and from its analysis they are able to tell what type of physical process gave rise to the observed radiation. Below we outline some typical processes which are considered by high energy astrophysicists.

1. Synchrotron Emission

This is emitted by an electric charge moving under the influence of a magnetic field. In Fig. E.1 we see an electric charge q moving in a circular orbit round a magnetic field H. The magnetic field does not affect the total energy of the electric charge. However, it changes the direction of motion of the charge and accelerates it. And an accelerated electric charge radiates electromagnetic waves. If the charge is moving slowly compared to the speed of light, the radiation emitted is of *cyclotron* type, encountered, for example, in the discussion of the infrared bursts in Chapter 6. If the charge is moving relativistically, i.e. with speeds approaching the speed of light, the radiation is of *synchrotron* type.

Synchrotron radiation is commonly used to describe radio emission in extragalactic sources and also optical emission in quasars. The characteristic frequency of radiation is given by

$$v \simeq 2.7 \, H \gamma^2 \left(\frac{m_e}{m} \right) \text{MHz},$$

where H is the magnetic field strength expressed in tesla perpendicular to the direction of motion of the electric charge. The factor γ, usually large compared to 1, is the relativistic factor (see Appendix B). We have assumed q to be the charge of the electron and for comparison m_e is the mass of the electron.

The charge in Fig. E.1 radiates in the forward direction of motion and the direction of the magnetic field in the electromagnetic wave is parallel to the direction of the original magnetic field H. The wave is thus polarized and this is a signature of the synchrotron process.

Fig. E.1

2. The Inverse Compton Effect

In Fig. E.2 we see an electron and a photon colliding. As a result of the collision, there is an exchange of energy and momentum between them and they go in different directions. This is the Compton process, first discovered in 1923 by A. H. Compton. In Compton's experiment the photon collided with a free electron in a material of low atomic weight, and as a result the photon *lost* energy and the electron *gained* it.

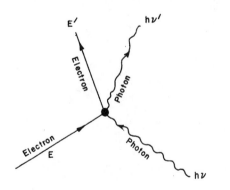

Inverse Compton Proces $\Rightarrow E' < E, h\nu'_1 > h\nu$
Compton Process $\Rightarrow E' > E, h\nu' < h\nu$

Fig. E.2

The astrophysicist is interested more in the reverse process in which a low energy photon colliding with a fast electron gains energy at the expense of the electron energy. It was this *inverse* Compton effect which was invoked in Chapter 7 to account for the X-ray halo of M 87. There the low energy microwave photon was changed into a high energy X-ray photon.

3. Bremmstrahlung

This German word means radiation produced by braking, and it describes a process which is commonly found to operate inside hot plasma. A *plasma* is a mixture of positively charged ions and negatively charged electrons. At high temperatures the electrons can no longer be kept bound to the nuclei and they acquire enough speeds to free themselves. The ion is an atom stripped of some of its outer electrons.

When such a pair made up of an electron and an ion collide the electron suffers a change of velocity and in this process it radiates (see Fig. E.3). The amount of radiation and its frequency depend on several factors including the plasma density and temperature. It is this process which is responsible for production of X-rays from heated accretion discs round compact objects.

4. Black Body Radiation

This is a general term used to describe the final equilibrium state of radiation which is essentially confined in a given volume. Normally, radiation has emitters and absorbers. What would happen to the confined

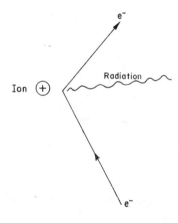

Fig. E.3

radiation if it is continually emitted and absorbed? It reaches a steady state in which there is a fixed number of photons in every given range of energies. Although individual photons may be emitted or absorbed, the overall statistical picture remains the same.

In this state the number of photons with energy in the range E to $E + dE$ is given by a function $N(E) \, dE$, where

$$N(E) = \frac{\alpha E^2}{e^{E/k\theta} - 1},$$

where α is a fixed constant. The distribution is characterized by the single parameter θ, the temperature of the black body, k being a constant known as the Boltzmann constant.

Black body radiation is useful in astrophysics in several contexts. It provides a first approximation to the radiation from stars. It also serves as a model for the radiation from the hot accretion discs emitting X-rays. The microwave background very closely resembles a black body spectrum.

The significance of black body radiation lies in the fact that it does not depend on the specific details of the radiation mechanism. It simply requires the radiation to settle down to an equilibrium state, which it does if it is effectively confined in a finite volume.

Table I

Catalogues of Astronomical Objects

As the techniques of observation improved, the number of astronomical surveys also increased and now there is a proliferation of catalogues of objects seen in different surveys. Below is given a partial list of such catalogues – those which have been referred to in the text. Although the method of listing a particular object in a catalogue is not universal, most modern catalogues list an object with the help of two coordinates* – right ascension and declination. Thus an object with the number $1221 + 114$ has the right ascension $12^h\ 21^m$ and declination $+11.4°$.

Stars	Bayer introduced a classification in which the brightest star in a constellation was designated by letter α, the next brightest by β, and so on. Thus α Cygni is the brightest star in the constellation of Cygnus.
NGC	The New General Catalogue of diffuse astronomical objects in the visual range of the electromagnetic spectrum. The catalogue was first published by J. L. E. Dreyer in 1888 and has since been augmented.
IC	Two Index Catalogues were published as supplements to the NGC catalogue, in 1895 and 1908.
M	The Messier catalogue of diffuse astronomical objects was first prepared by C. Messier in 1781.
HD	Star catalogue published by the Harvard College Observatory and named after the astronomer Henry Draper.
HDE	The extension of the HD catalogue.
3C, 4C	Catalogues of extragalactic radio sources prepared by the radio astronomers in Cambridge. Thus 3C is the third Cambridge catalogue, 4C the fourth Cambridge catalogue, and so on.
OQ, OX	The Ohio catalogues of radio sources list sources in the series OB to OZ (omitting OO). Each letter of the alphabet in the second place corresponds to an hour of right ascension. Thus a source listed as OQ has right ascension in the 14^h–15^h range. The numbers following the letter give further information about

---

* See *Glossary* for the meaning of these coordinates.

Table 1 Catalogues of Astronomical Objects 203

the coordinates of the source. Thus OQ172 has declination in the range $10°-20°$ (corresponding to number 1) and has the right ascension 14^h. 72.

CP   Denotes Cambridge Pulsar, i.e. a pulsar found through a survey by the radio astronomy group in Cambridge.

PSR   The more commonly used label for a pulsar. The numbers following it list the right ascension and declination of the object.

Cyg X, Her X, etc. The original listing of X-ray sources by the constellation in which they were found. Thus Cyg X-1 is the brightest source in the constellation of Cygnus.

3U   The first systematic catalogue of X-ray sources was provided by the survey taken by the UHURU satellite. Sources with the prefix 3U appear in the third UHURU catalogue.

MXB   Catalogue of X-ray bursters prepared by the X-ray astronomers at M.I.T.

Table II

Fundamental Constants of Physics and Astrophysics

Speed of light	$c = 2.9979250 \times 10^8 \,\mathrm{m\,s^{-1}}$
Planck's constant	$h = 6.62620 \times 10^{-34} \,\mathrm{J\,s}$
	$\hbar = 1.05459 \times 10^{-34} \,\mathrm{J\,s}$
Gravitational constant	$G = 6.670 \times 10^{-13} \,\mathrm{N\,m^2\,kg^{-2}}$
Electron charge	$e = 4.80325 \times 10^{-10} \,\mathrm{esu}$
Electron mass	$m_e = 9.10956 \times 10^{-31} \,\mathrm{kg}$
Fine structure constant	$e^2/\hbar c = 1/137.036$
Classical electron radius	$e^2/m_e c^2 = 2.81794 \times 10^{-15} \,\mathrm{m}$
Proton mass	$m_p = 1836.11 \, m_e$
Compton wavelength of electron	$h/m_e c = 2.426310 \times 10^{-12} \,\mathrm{m}$
Electronvolt	$1\,\mathrm{eV} = 1.602192 \times 10^{-19} \,\mathrm{J}$
Frequency associated with 1 electronvolt	$\nu_0 = 2.417965 \times 10^{14} \,\mathrm{s^{-1}}$
Temperature associated with 1 electronvolt	$\theta_0 = 11\,604.8 \,\mathrm{K}$
Boltzmann constant	$k = 1.38062 \times 10^{-23} \,\mathrm{J\,K^{-1}}$
Radiation density constant	$a = 7.56464 \times 10^{-16} \,\mathrm{J\,m^{-3}\,K^{-4}}$
Unit for nuclear cross-section (Barn)	$1\,\mathrm{B} = 10^{-28} \,\mathrm{m^2}$
Energy released by fusion of 4 H nuclei into a He nucleus	$26.72 \,\mathrm{MeV}$
Astronomical unit	$1\,\mathrm{AU} = 1.495979 \times 10^{11} \,\mathrm{m}$
Parsec	$1\,\mathrm{pc} = 3.085678 \times 10^{16} \,\mathrm{m}$
	$= 3.261633 \text{ light years}$
Light year	$1\,\mathrm{l.y.} = 9.460530 \times 10^{15} \,\mathrm{m}$
Solar mass	$M_\odot = 1.989 \times 10^{30} \,\mathrm{kg}$
Earth mass	$M_\oplus = 5.976 \times 10^{24} \,\mathrm{kg}$
Hubble's constant	$H \sim (16\text{–}32)\,\mathrm{km\,s^{-1}}$ per million l.y.
	$H^{-1} \sim (0.9\text{–}1.8) \times 10^{10} \text{ years}$

Glossary

absolute magnitude See *magnitude.*

absolute space When formulating his laws of motion Newton had to specify a frame of reference against which the velocities and accelerations of moving bodies are to be measured. Newton postulated such a background and called it the absolute space. (See also *inertial frame.*)

acceleration The rate of change of velocity. The change of velocity can occur in two ways – through a change of speed and through a change of direction of motion.

accretion disc Matter from the surroundings attracted by a rotating massive object or by a compact member of a binary star system settles down in the form of a disc around the object. This is known as the accretion disc.

alpha particle The nucleus of helium, containing two neutrons and two protons. In *alpha decay* a heavy nucleus breaks up and emits alpha particles.

Ångstrom (Å) A unit of length equal to 10^{-10} m. This unit has been found suitable for measuring the wavelength of light waves.

apparent magnitude See *magnitude.*

astrophysics Subject dealing with the physical properties of astronomical objects.

atomic number The total number of protons in an atom. This is also called the *charge number* (see Appendix D).

Atomic weight denotes the total number of neutrons and protons in the atom. It is also called the *mass number.*

baryons See Appendix D.

beta decay In beta decay a neutron decays into a proton, giving out an electron, an anti-neutrino, and some energy. In general, *beta processes* in a system of neutrons and protons lead to a stabler system through changes of neutrons to protons or vice versa.

binary stars Two stars moving round each other under their mutual gravitational force of attraction.

black hole An object with such a strong gravitational field that not even light can escape from its surface. According to conventional ideas a black hole may form when a massive object undergoes uncontrolled contraction (a collapse) because of the inward pull of its own gravity.

blue-shift If the spectrum of an object undergoes a systematic increase of frequency, it is said to be blue-shifted. In the visual range such a shift will result in moving the spectrum towards the 'blue' end. Hence the name 'blue-shift'.

burster An object emitting radiation in short bursts. X-ray astronomy has revealed the existence of several bursters.

centrifugal force The radially outward force experienced by a body moving in a circular path about a fixed point as the centre.

cosmic rays Flux of high energy particles from outer space. Cosmic rays may originate in supernova explosions and in pulsars.

cosmology Subject dealing with the large scale structure and evolution of the Universe. The *cosmological principle* is a simplifying assumption in cosmology which states that the Universe looks the same in all directions and at all points when viewed from special vantage points at any given epoch. The special vantage points are idealized positions which in the real Universe are approximated to by galaxies.

declination and right ascension These are equivalent to the latitude and longitude coordinates, but measured for astronomical objects, and are defined in this way. Imagine the Earth's equator to be projected across the sky, with the polar axis similarly extended both ways to give imaginary north and south poles on the sky. The *declination* of an object is its latitude with respect to this projected equator (called the *celestial equator*). It is positive if it is located north of the equator and negative if it is located south of it. The *right ascension* is measured along the celestial equator (like the longitude), from a point where the sun's path projected across the sky meets the equator at the spring equinox. (The sun's projected path is another circle called the *ecliptic* and it meets the celestial equator in two points, once in spring and once in the autumn.)

Deuterium Heavy hydrogen. The nucleus of ordinary hydrogen contains only one proton and no neutrons while the nucleus of deuterium contains a proton and a neutron.

Doppler shift See Appendix C.

eccentricity See Fig. 2.3.

entropy A mathematical measure of the disorder of a physical system. The law of increase of entropy implies that a physical system passes through succesive states of increasing disorder.

Escape speed The minimum speed needed to fire a projectile from the surface of a massive object for the projectile to be able to escape to infinity. If the projectile has speed lower than the escape speed, the gravitational attraction of the body traps it and keeps it within a finite distance from the body.

Euclidean geometry Geometry based on Euclid's postulates.

event horizon Boundary of a space-time region such that events taking place beyond it will never be visible to a given observer in the region.

fission The breakup of a heavy atomic nucleus. This happens when the force of nuclear attraction binding the nuclear particles together becomes weaker than the electrical forces of repulsion.

flux The flux of radiation measures the amount of radiant energy crossing in unit time a unit area held perpendicular to the radiation beam.

frame of reference The position and timing of an event can be specified relative to landmarks in space and time. These constitute the frame of reference. Usually the specification is done with numbers called the *coordinates*. For example, the latitude and the longitude are coordinates specifying the position of a terrestrial event.

frequency The number of times a repetitive event occurs in a unit of time is called the frequency of the event. In wave motion the frequency refers to the number of oscillations per second. For example, in an electromagnetic wave the strength of the electric field oscillates and the frequency of oscillation can be measured by examining the up and down motion of an electric charge kept in the path of the wave.

fusion The process in nuclear physics which brings light nuclei together to form a bigger nucleus.

gamma (γ) rays Radiation of very high frequency, exceeding 10^{20} Hz.

globular cluster A large cluster of stars nearly spherical in shape. A typical globular cluster may contain 100 000 to ten million stars moving in one another's gravitational attraction. In our Galaxy there are about 100 such clusters.

gravitational collapse A state of uncontrolled contraction of a massive object under its own force of gravity. At this stage no other physical forces are capable of providing resistance to contraction. According to Einstein's general relativity the ultimate state of the collapse is a singularity of space and time (see *singularity*).

half-life The characteristic time of decay of a radioactive nucleus. This is the time over which half the total population of such radioactive nuclei are found to have decayed.

hertz (Hz) Unit of frequency named after the nineteenth century scientist H. Hertz who first produced and detected electromagnetic waves in the laboratory.

homogeneous Having the same physical properties all over space.

Hubble's constant The ratio of the apparent velocity of recession of a distant galaxy to its distance from us. Its present estimate is in the range 16–32 km s^{-1} per million light years.

inertia The property of matter characterizing its resistance to change of state of rest or of uniform motion in a straight line.

inertial frame Frame of reference in which Newton's first law of motion holds good, i.e. a frame of reference in which a body under no forces remains at rest or has uniform motion in a straight line. All inertial frames are in uniform motion in a straight line relative to one another. Newton's absolute space is one of this system of inertial frames.

ion An atom or a molecule is electrically neutral. When, however, it is stripped of one or more of its electrons it acquires a positive charge. It is then called a positive ion. A process in which an assembly of atoms (or molecules) is converted to ions is called *ionization*.

isotopes Atomic nuclei with the same number of protons but with a different number of neutrons. (See Appendix D.)

isotopic anomaly An unexpectedly different ratio of isotopes found in a piece of matter. See Chapter 5 for a description of how isotopic anomalies have helped in determining the age of a meteorite.

isotropy Having the same property in all directions.

lambda (λ) *force* A force of repulsion introduced by Einstein in his equations of general relativity in order to obtain a model of the static Universe. The repulsive force between any two lumps of matter, according to Einstein's law, was supposed to be λ times the distance between them. The constant λ was so small that this force was negligible in the context of the gravitational forces acting within the solar system: it produced a significant effect only on the cosmological scale. Later, when Hubble's observations (see Chapter 8) became known, the model of a static universe became unsustainable and Einstein dropped the λ-term from his equations. A few cosmologists still think that such a term is necessary to account for the present cosmological data.

light cone Region in space-time accessible to light rays emitted from a given point.

light year Distance travelled by light in one year. This is approximately 9 million million kilometres.

luminosity This is the amount of energy emitted by a given object in a unit of time. It is therefore like the rating of a power station. The sun's luminosity is 4×10^{26} joules per second.

Mach's principle A hypothesis advanced in the late nineteenth century by the scientist–philosopher Ernst Mach. Mach argued that the *inertia* of a body is not an intrinsic property of the body; rather it is the consequence of the body's presence in an otherwise non-empty universe. Mach gave perspasive arguments for this hypothesis (see Chapter 9) but gave no mathematical theory for it.

magnitude An astronomical scale quantifying the brightness of heavenly bodies. There are two types of magnitudes: the *absolute* magnitude compares the brightnesses of different objects as if they were viewed from the same distance, while the *apparent* magnitude compares the brightnesses of different objects as seen from the Earth. Magnitudes are measured on a logarithmic scale with equal intervals of the magnitude scale corresponding to equal *ratios* of brightness. Thus a factor of 100 in brightness is represented on the magnitude scale by an interval of 5: a star of the 1st magnitude is 100 times brighter than a star of the 6th magnitude. The magnitude scales are normalized such that the sun has an apparent magnitude of − 26.82 and an absolute magnitude of 4.75 (representing the *total* or bolometric brightness of the sun).

main sequence The phase in a star's life when it is generating energy mainly by the fusion of hydrogen to helium.

meteorite Small sized bodies (ranging in diameter from a few centimetres to several metres) which move in the solar system under the gravitational pull of the sun and the planets. The chemical composition of meteorites provides valuable clues to the origin of the solar system.

microwave background Isotropic background of radiation mainly in the microwaves, first discovered in 1965. The spectrum of the radiation is approximately that of a black body radiation of temperature 3 K.

neutrino An elementary particle of zero electric charge travelling with the speed of light. It is denoted by the Greek letter v.

neutron star Star composed of matter mainly in the form of neutrons. The density of matter in the neutron star may be at least ten million times the density of water and can be as high as a thousand million million times the density of water.

non-Euclidean geometry A geometry based on axioms differing from Euclid's in one or more respects.

nucleus Central region. This word may be used in the context of an atom or of a galaxy. An atomic nucleus has a characteristic size of 10^{-14} m while a galactic nucleus may be hundreds of light years across.

parsec An astronomical measure of distance (see Table II), a little more than three light years. It is the distance of a star at which the radius of the Earth's orbit round the sun would subtend an angle of one arc second.

perihelion The point in the orbit of a planet where it is closest to the sun.

photon The particle associated with light. Light appears to have a dual nature. It behaves like a wave and also a swarm of particles. Max Planck in the year 1900 suggested that a beam of light of specified frequency v (say) is made up of discrete packets of energy, each packet having energy hv, where h is a universal constant known as Planck's constant. Photon is the name given to a typical light packet.

Planck's constant See *photon*.

Planckian curve The curve describing the variation of intensity with frequency in a black body radiation (see Appendix E).

plasma Gas containing ionized atoms or molecules and free electrons (see *ion*).

polarization A property of the light wave which tells how the electric and magnetic disturbances in it are aligned. In a *plane polarized* wave the electric disturbance is in one plane while the magnetic disturbance is in a perpendicular plane, the wave itself propagating along the direction parallel to the line of intersection of these planes.

proper time Time measured by an observer in his own rest frame.

pulsar Astronomical objects emitting highly regular pulses of radiation with time-scales of the order of a second. It is believed that pulsars are rotating magnetized neutron stars. (see Chapter 5 for further details.)

quantum physics Physics based on quantum theory. This theory imposes certain fundamental limitations on measurements of physical quantities and leads in many cases to a discrete (i.e. jumpy) behaviour of matter where classical (pre-quantum) physics predicted a continuous behaviour. Experiments in atomic physics have confirmed the validity of quantum theory.

quasar (QSO) Compact extragalactic object which presents a star-like appearance in spite of being a much more massive and powerful radiator than a star. (*QSO* stands for quasi stellar object.) A few per cent of quasars emit radio waves, while the recent X-ray observations have shown that quasars are very often emitters of X-rays.

radiogalaxies Galaxies emitting radio waves.

red-shift A light wave is said to be red-shifted if its wavelength increases between its point of emission and point of reception. The fractional increase in wavelength (i.e. the net increase divided by the original wavelength) is called the red-shift. Astronomers know of three possible causes of red-shift. (1) The *Doppler red-shift* arises from the fact that there is a relative motion of the source of waves away from the observer. (2) The *gravitational red-shift* describes the increase in wavelength when light travels from a region of strong gravity to a region of weak gravity. (3) The *cosmological red-shift* arises when light travels in an expanding universe. While one of these explanations has proved sufficient to explain the observed red-shift of an astronomical object, there are some indications that the large red-shifts of quasars may not arise through any of these three causes. This has led to the phenomenon of *anomalous red-shifts* discussed in Chapter 9.

relativity See Chapters 3 and 4.

right ascension See *declination and right ascension*.

singularity This word in general indicates something unusual. To the relativist the word singularity implies the breakdown of the mathematical rules defining the geometry of space-time. According to Einstein's equations, it seems that in the presence of matter behaving according to the conventional laws of physics the space-time must at some stage have a singularity. Specific instances of singularity are the so-called big bang origin of the Universe and the ultimate state of collapse of a massive object. In the latter case the singularity is hidden from the outside observer by a horizon (see *event horizon*). A space-time singularity which is not so hidden is called a *naked* singularity.

sink and source Concepts in thermodynamics. Source is where energy originates and sink is where it is deposited. These concepts are mainly used for transfer of heat from one place to another. The laws of thermodynamics place certain limits on how if at all the energy transfer is to take place.

space-time The three dimensions of space and one dimension of time together form a single four dimensional entity called space-time. Einstein's relativity theory first brought home to physicists the point that space and time are closely linked and that it is more convenient to think of the laws of physics in the unified space-time than in the separately compartmentalized space and time.

space-time diagram A geometrical way of depicting events in space-time is to think of them as points in a diagram describing the three dimensions of space and one of time. Thus the point describes *where* the event took place and *when* it took place.

spectrum Although in its simplest form a light wave is characterized by its frequency, typical light from an astronomical object is usually a mixture of waves of different frequencies. In a spectrum of the object these waves are separated out. Thus the spectrum of sunlight separates out the waves, which manifest themselves as different colours ranging from the red to violet, each wavelength corresponding to a colour shade. Over and above a continuous range of wavelengths, the spectrum may also show dark and bright lines. A dark line arises from the absorption of light by some absorbing material *en route*. This material, usually in the form of atoms or molecules, absorbs only waves of certain fixed frequencies.

The rules of quantum theory (see *quantum physics*) were able to explain these *absorption lines* in terms of 'jumps' which take place when an electron in an atom, say, makes a transition to a state of higher energy. Since the energy levels to which the electron can jump are a discrete set, waves of only specified frequencies are absorbed by the atom. Hence dark lines appear in the spectrum of the object at these frequencies. The bright line or the *emission* line arises when the electron in a hot atom 'jumps down' to a state of lower energy, thereby releasing a light wave of specified frequency.

surface gravity A quantity denoting the acceleration felt by a test particle near the surface of the object. For a stationary black hole the surface gravity also signifies its temperature, i.e. its ability to radiate to its surroundings.

synchrotron radiation See Appendix E.

temperature An isolated physical system tends to reach a state of equilibrium when, as a result of exchanges of energy between its various constituents, there is an equipartition of energy between the various components. At this stage the temperature is a measure of average energy per component. In the so-called *absolute* scale of temperature, the temperature increases in proportion to the average energy per component; the state of absolute zero temperature signifying the state of zero energy.

thermodynamics Science dealing with exchange of heat energy and its relation to work and the mechanical behaviour of physical systems.

thermonuclear reactions Reactions which take place when different atomic nuclei are brought together at high temperatures. In these reactions energy is absorbed or released and changes in the structure of participating nuclei take place.

unified theory Einstein attempted a theory, called the unified theory which, he hoped, would encompass all physical interactions. At that time the unified theory was supposed to include gravity and electromagnetism. The attempt did not succeed and later it became clear that a unified theory must also include the strong and weak interactions (see Appendix D). Recently a *unified gauge theory* has successfully linked electromagnetism with the weak interaction and it is hoped that some day a *grand unified theory* will also bring the strong interaction into the fold. Gravity, however, remains aloof from these attempts at unification.

white dwarf A star of highly dense matter in the form of electrons and nuclei. The density of matter in a white dwarf may be as high as a hundred million times the density of water, the matter being in degenerate form (see Box 5.1).

white hole An exploding object apparently emerging from a space-time singularity. In a certain sense a white hole is the time reversed version of a black hole, which is an object collapsing into a singularity. Unlike the black hole, a white hole is a bright object.

X-ray astronomy Astronomy based on the X-ray part of the electromagnetic spectrum (see Appendix A).

Further Reading

Cosmology + 1, readings from *Scientific American* with introduction by O. Gingerich (W. H. Freeman, San Franscisco, 1977).

Gravity, by G. Gamow (Heinemann, London, 1962).

The Physics Astronomy Frontier, by F. Hoyle and J. V. Narlikar (W. H. Freeman, San Francisco, 1980).

The Structure of the Universe, by J. V. Narlikar (Oxford University Press, Oxford, 1977).

The State of the Universe, ed. G. T. Bath, a collection of lectures by several speakers (Oxford University Press, Oxford, 1980).

The Big Bang, by J. Silk (W. H. Freeman, San Francisco, 1979).

The First Three Minutes, by S. Weinberg (Deutsch, London, 1977).

Index

absolute space 165, 166
absolute time 189
absorption lines 186, 187
acceleration 15
accretion discs 105–8, 113, 200, 201
accretion on to a black hole 131–2
Adams, J. C. 25
aether 189
age of the Universe 146–7
Airy 25
^{26}Al 85–6
Alfven, H. 197
Allende meteorite 85–7
alpha particle 77, 195
Alpher, R. 150
angular momentum of a black hole 59–70
anomalous red-shifts 176–83
ANS satellite 113
anti-matter 197
anti-particle 197
Apparao, K. M. V. 116–18, 175
Aristotle 15
Arp, H. C. 178, 181, 183
Ashok, N. M. 116, 117
Assousa, G. E. 84

Baade, Walter 7, 83, 127
Bahcall, J. N. 112, 114, 176, 177
baryon 196
Bell, Jocelyn 3
bending of light 46–7
Berkeley, Bishop 166
Bertotti, B. 167
beta decay 196
Bethe, H. 150
big bang 10–11, 145–7, 184
binary pulsar 93
binary star system 4, 5, 101–8, 111
binding energy 30, 31, 195

black body 72, 150, 155, 200–201
black hole 32, 33, 56–75, 89, 95
  analogy with thermodynamics 67–74
  area of 61, 64–7
  as a black body 72
  Cygnus X-1 108–9
  dynamics of 101–2
  energy extraction from 64–7
  irreducible mass of 65
  Kerr 60, 64–5
  Kerr–Newman 60, 64–70
  massive 113–14
  mini, primordial 73, 158
  Newtonian v. relativistic 57–8
  physics of 63–70
  Reissner–Nordström 60
  Schwarzschild 57, 60, 64–7, 73
  stationary axisymmetric 68–70
  supermassive 123–6, 130–4
  surface gravity of 68–73
Blandford, Roger 131, 134
blue-shift 175, 180, 183, 194
Boltzmann 67
  constant 72, 154
Boksenberg, A. 123
Bolyai 40
Bondi, Hermann 104, 105, 180
Braes, L. 108
Brahe, Tycho 3
Brecher, Kenneth 2
bremmstrahlung 200
Burbidge, E. M. 177
Burbidge, G. R. 128, 177

3C 273 132, 137, 180
Canezares, C. R. 114
Canis Major R1 84–5, 87
Carr, B. 157, 158
Carswell, R. F. 137

Cavaliere, A. 130
CCD 123
celestial mechanics 23–6
Challis 25
Chandrasekhar, S. 53
Chitre, S. M. 116–18, 136, 157
Chiu, H. Y. 158
Chubb, T. 97
Clark, G. W. 97, 112
Clausius 67
closed Universe 145, 146
comets 23, 24
Compton, A. H. 199
conservation of energy 28
Cooke, D. J. 91
cosmic censorship 75
cosmological principle 145
cosmology and particle physics 158–61
Crab Nebula 1–3, 76, 82, 90, 92, 98, 120
Crab pulsar 90, 92, 98–100
curvature of space-time 43–5
cyclotron emission 116, 198
Cygnus A 7, 8, 127–30
Cygnus X-1 4, 99, 100, 102, 108–9

Dadhich, N. 175
Das, P. K. 181, 183
degenerate pressure 87, 88
de Sitter, W. 142
de Sitter universe 142
deuterium 152–4
Dicke, R. H. 154, 167
Dirac, P. A. M. 197
Doppler effect 55, 101, 125, 140, 175, 183, 193–4
Duncan, M. J. 126
Dyson, Sir Frank 47

Earth's rotation 164, 166
Eddington, Sir Arthur 46–8, 107
  luminosity 106–8, 131, 134
Edmunds, M. J. 157
Einstein, Albert 13, 37, 141, 142, 167, 189–91
  Universe 141–2
Einstein–de Sitter model 172
Einstein Observatory 98, 108, 114, 122, 133, 137, 138
electromagnetic interaction 11

electromagnetic wave 12, 185–6, 189
emission lines 186, 187
entropy 67
ergosphere 61, 62, 64–7
escape speed 32, 52, 82
Euclid 39
Euclidean geometry 39
event horizon 57, 60
Ewen, H. L. 127
expanding universe 10, 139–45
explosive nucleosynthesis 82, 86

Fabian, A. 111
faster than light motion 135–7
Foucault pendulum 166
Fowler, L. A. 94
Fowler, W. A. 123, 130, 151, 180
Friedman, H. 97–8, 120
Friedmann, A. 142
  models 145–7, 169–72, 184

galaxy 6–8, 119, 127, 139–41, 155, 180
galaxies
  active 120
  colliding 127–30
  elliptical 119, 120
  spiral 119, 120
Galileo Galilei 12, 15
Galle, J. G. 25
gamma ray 4, 5, 187
  bursts 4–5, 73
Gamow, George 149–50, 154, 158
Gauss 40
Gellmann 196
general relativity 13, 44–51, 141, 175, 184
  experimental tests of 46–51
geodesic 44
geometry of space and time 39–42
Geroch, Robert 75
Giacconi, R. 97
Glashow, S. L. 158, 197
globular clusters 5–6, 110
  X-ray sources in 111–12
gluon 196
Gold, T. 91
Grand Unified Theory (GUT) 159, 197
gravitation, gravity 11–13
  constant of 23, 27
  energy of 26–31, 129

law of 23
permanence of 39
strength of 31–5, 52
gravitational collapse 35, 53–8
end point of 74–5
gravitational blue-shift 49, 74
gravitational red-shift 49, 54, 55, 180–1
gravitational radiation 93–6
graviton 59
Grindley, E. 113, 114
Gunn, J. E. 137, 176
Gursky, H. 97, 113
gyroscope precession 62

hadron 196
Halley, Edmond 23
Halley's comet 23, 24
Hartwick, F. D. A. 123
Hawking, Stephen 70, 75
Hawking process 69, 70–4, 157
Hayashi, Chushiro 77, 150
Hazard, Cyril 9
HDE 226868 108
helium 152–3
Helmholtz, Hermann 31
Herbst, W. 84
Hercules X-1 99, 102
Hermann, R. 150
Herschel, William 25
Hertz, H. R. 126
Hewish, Antony 3
Hey, J. S. 7, 127
Hills, J. 111
Hjellming, H. 108
homogeneity 142
Hooke, Robert 23
Hoyle, Fred 82, 104, 105, 122, 123, 130, 151, 171, 180
Hubble, Edwin 10, 139, 140, 144, 169, 176
Hubble's constant 141, 144, 147
Hubble's law 135, 141, 144, 176
Hulse, R. A. 93
Humason, M. 139, 140
HZ Hercules 100, 102

inertia 15, 164
law of 15
and cosmology 164–7
inertial forces 165–6

inertial observers 38, 39, 188, 189, 191
infrared emission 77
in bursts 116–18
inverse Compton process 121–2, 199–200
inverse square law 18–23, 38
ion 200
IPCS 125
isotopes 85, 195
isotopic anomalies 85
isotropy 142

Jansky, Karl 126, 127
Jones, A. W. 118

Kelvin, Lord 31, 67
Kembhavi, Ajit 171, 172
Kepler, Johannes 3, 12, 18
Kepler's laws 18–21, 101
Kerr, R. P. 60
kinetic energy 28, 191
King, I. 124
King models 124
Kleinmann, D. E. 115
Kleinmann, S. G. 115
Kristian, J. 123, 137
Kulkarni, P. V. 116, 117

λ-force 142
Lamb, F. 114
Landau 89
Landauer, F. P. 123
de Laplace, Pierre Simon 24, 33, 57
laser ranging of the Moon 51
lepton 196
Le Verrier, U. J. J. 25, 26
Lewin, W. 115
Lieber, Elinor and Alfred 2
light cone 191
Liller 1 115
Liller, W. 115
Lobatchevsky 40
Lorentz 190
Lorentz transformation 190
Lynden-Bell, D. 123, 167
Lynds, C. R. 123
Lyttleton, Raymond 104, 105

M 87 7, 120–6, 200
Mach, Ernst 166, 167
Mach's principle 167, 184

main sequence 77
Margo, C. S. 118
Maxwell, James Clerk 37, 185
McCullogh 94
Mercury 25, 26
meson 196
Messier catalogue 7, 120
Messier, Charles 120
Michelson–Morley experiment 189
microwave background 122, 154–8, 201
    from intergalactic dust 157
    from supermassive stars 157
Miley, G. 108
Mills, B. Y. 128
minibangs 175
Minkowski, Rudolf 7, 127
Mitchell, John 32, 57
Morrison, Philip 130
motion, laws of 14–18
Muller, C. A. 127
Munoz, M. P. 118
MXB 1730–333 115

naked singularity 75
Neptune, discovery of 25
neutrino 80, 149, 161
    emission from stars 80
    massive 149
    types 161
neutron stars 87–90, 107, 114, 116, 180
New General Catalogue 6
Newman, E. T. 60
Newton, Isaac 12, 14, 15, 38
Newton's bucket experiment 164–5
non-Euclidean geometries 40–2, 55, 61, 142, 169
Nordström, G. 60
Novikov, I. 151
nuclear fission 196
nuclear fusion 77, 196
null geodesic 47

O'Dell, S. 177
Oke, J. B. 137
Oort, J. H. 127, 139
open Universe 145, 146
Opik, J. 82
Oppenheimer 89
optical identification 9, 100, 108, 127
OQ 172 132

origin of Solar System 82
Orion Nebula 77
Ostriker, J. P. 112, 114
OX 169 134

Paolini, F. 97
parallel postulate 40–1
Parsons, S. J. 7, 127
Pati, J. 161
Pauli's exclusion principle 88
Peebles, P. J. E. 151, 154
Penrose, Roger 64, 75
Penrose process 64–6
perihelion 19
    of binary system 95
    of Mercury 26, 45
Penzias, A. A. 122, 155
Phillips, J. W. 7, 127
photon 46, 50, 59, 72, 121, 122, 169, 181, 186
photon to baryon ratio 158
pion 196
Planck, Max 186
Planck's constant 50, 72, 168, 187
plasma 200
polarization 128, 185
Price, R. H. 59
Price's theorem 59–60
primordial nucleosynthesis 147–54
Principia 15, 23, 166
Pringle, J. 111
Proctor, R. A. 140
proto stars 77
pulsar 3–4, 90–3
pulse window 90
Purcell, E. M. 127

quantum theory 167
quarks 160–1, 196
quarko synthesis 161
quasi stellar object, quasar 9, 132–8, 199
    anomalous red-shifts of 176–83
    energy production in 133–4
    faster than ling motion in 135–7
    images by gravitational bending 136, 137

Racine, R. 84
radar echo delay 50–1
Radhakrishnan, V. 91

radio sources 7–8, 128–2
Ramadurai, S. 157
Rana, N. C. 157
Rapid Burster 99, 114, 115–18
Reber, G. 127
red giants 78
red-shift 9, 49, 54–6, 74, 132, 133, 140
  of fourth kind 181–3
Rees, M. J. 111, 131–2, 136, 157
Reissner, H. 60
Richards, P. L. 155
Riemann 40
Roche, E. 103
Roche lobe 103, 106
Roll, P. G. 154
Ryle, Martin 128

Salam, A. 158, 197
Sargent, W. L. W. 123
SAS-3 satellite 113, 115
scale invariance 167–9
scale invariant theory of gravity 169–72, 181–3, 184
Schmidt, Maarten 9, 176, 180
Schwarzschild, K. 44, 57
  barrier 57
  black hole 57
Sciama, D. W. 167
Sco X-1 98
Seyfert, C. 7
Seyfert galaxy 7
Shapley, H. 139
shock wave 81, 84, 88
Shortridge, K. 123
Silk, J. 112
SIT 123
Smith, F. G. 7, 127
Snyder 89
Soloman, P. 177
space telescope 137
space-time diagram 43
space-time singularity 75, 147, 171, 174
special relativity 37, 38, 39, 44, 135, 136, 167, 188–92
speed of light 38, 167, 185, 189
spin of an interaction 59
spinar 130, 132
spin–orbit coupling 96
Spitzer, Lyman 8
star formation 77, 82–7

starquake 93
static limit 61–3
stellar evolution 76–8
Stockton, A. 176, 177, 178
Strittmatter, P. S. 177
strong interaction 11, 196
Sun 7, 12
supermassive black hole 123–6, 130–2, 133–4, 137
supernova 3, 76, 78–87
  core bounce 88
  induced star formation 84
  light curve 82, 83
  remnant 92–3, 99
  shock wave 81, 84, 88
synchrotron emission 92, 128, 129, 198

Tayler, R. J. 151
Taylor, J. H. 93, 94
temperature 67
thermal bremmstrahlung 122
thermal equilibrium 68
tidal interaction 74
time dilatation 191, 192
tunnel effect 73
twin exhaust model 131–2

3U 1820–30 113
UHURU satellite 4, 98, 99
unified gauge theory 197
Uranus, orbit of 25

van de Hulst, H. C. 127
variation of mass 168–72
Vela pulsar 92
velocity dispersion 125
Virgo cluster 120
Vulcan 26

Wade, C. 108
Wagoner, R. V. 151
Walsh, D. 137
Warner, J. 84
weak interaction 11, 196
Weber, J. 95
Weinberg, S. 158, 197
Westphal, J. A. 123, 137
Weymann, R. J. 137
Wheeler, J. A. 60
Wheeler, J. C. 126
white dwarfs 88, 111, 180

white holes 173–6
Wickramasinghe, N. C. 157
Wilkinson, D. T. 154
Wilson, P. 123
Wilson, R. 122, 155
Woody, D. P. 155
world line 43
Wright, E. L. 115

X-ray astronomy 97–8
X-rays 187
  spectroscopy 122
X-ray sources (types)

binary 99–104, 106, 107
bursts 99, 110–14
clusters of galaxies 122
Rapid Burster 99, 114, 115–18
supernova remnants 99

Young, P. J. 123, 137

Zeldovich, Ya. 150, 158
zero mass epoch 170, 171, 173
zero mass surfaces 173
Znajek, R. L. 134
Zweig 196